Plant Diseases

Plant Diseases

Gyan Deep Singh

Oxford Book Company
Jaipur ▪ New Delhi

Published By

Oxford Book Company

Regd. Off. "Bony Residency", Gate No. 2
Opp. Tilak Public School, Vishweswariya Nagar,
Gopalpura Road, Jaipur - 302018, India
Ph: 0141 - 2761280, Telefax: 2761381
email - oxfordbookjpr@gmail.com
website - oxfordbookcompany.com

Delhi Office: 102, 1st Floor, Satyam House,
4327/3, Ansari Road, Darya Ganj,
New Delhi - 110002, India
Ph: 011-45652440

Published in India, 2025

ISBN : 9789350300008

Price : ₹ 2500/-

Typeset by Shivangi Computers, Jaipur

Printed and bound in India by Thomson Press Ltd.,
New Delhi

Preface

Plant Diseases is a complete book on plant diseases and their management. Dynamic advances have been made in the last century in plant pathology leading to vast economic benefits. The topics cover biotechnological management of viral diseases, post-harvest diseases of perishables, diseases of seed spices, legumes and mustard. Special chapters on fungi and superficial mycoses, insect and fungal galls and molecular approach for host-pathogen interactions have been included.

Major nematode diseases in plants and their management with reference to India have been discussed. Innovative and novel approaches for the management of plant pests have been described which could be successfully used on a commercial level. India has made rapid strides in the field of agricultural research and education. Green revolution witnessed the use of plant protection umbrella for new varieties and hybrids.

Plant diseases are reasonably uniform in size and stability. Understanding and appreciation of these aspects of the science are essential for the professional plant pathologist, but in a terminal course primarily for agricultural students the practical aspects of the science must receive first consideration. The arrangement of disease within chapters is a compromise in which the systematic relationship of pathogens are given first consideration, but with such modification as is necessary to avoid too abrupt and illogical transitions between disease types.

Prof. Gyan Deep Singh

Contents

Chapter 1

Introduction

The study of plant diseases is known as plant pathology. Infectious diseases are caused by living organisms called pathogens. Noninfectious diseases caused by environmental stress and damage by weather and other environmental factors also will be covered. Indirectly, environmental factors that cause a plant to be stressed may result in the plant's gradual decline. Decline results in the plant being more susceptible to disease organisms.

Because of this, diagnosing plant diseases can be tricky. The real cause of a problem may be the stress factors, with the disease simply being a secondary factor.A plant disease is broadly defined as any condition in which a plant differs in some way from a normal plant in either structure or function. Plant diseases have had an important role in history. The holy fire in the Bible is believed to have been due to the ergot fungus in the heads of grain which made bread poisonous. The great famine in Ireland in the mid-1800s was caused by the destruction of the potato crop by late blight.

WHAT CAUSES PLANT DISEASE?

Plant diseases are generally divided into two groups based on their cause. Nonparasitic diseases are induced by some genetic or environmental factor such as nutrient deficiencies, extreme cold or heat, toxic chemicals (air pollutants, weed killers, or too much fertilizer), mechanical injury, or lack of water. These diseases cannot be transmitted to healthy plants and their control depends solely on correcting the condition causing the disease.

Parasitic diseases are caused by living organisms which derive their food by growing as parasites upon other plants. The most common causes of parasitic diseases are fungi, bacteria, viruses, and nematodes. A few seed-producing plants, such as the mistletoes, can also cause plant diseases. Fungi are plants that lack the green colouring (chlorophyll) found in seed-producing plants and therefore cannot manufacture their own food. There are between 50,000 and 100,000 different species of fungi of many types and sizes, but not all are harmful. Most are microscopic in size, but some, such as the mushrooms, are quite large. Fungi reproduce by spores. Bacteria are very small, one-celled plants that reproduce by simple fission. They divide into two equal halves, each of which becomes a fully developed bacterium. Plant diseases are frequently given their common names based on the plant part affected, by the way the plant is affected, or by the cause of the disease.

Viruses are so small they cannot be seen with an ordinary microscope. Many of the viruses that cause plant diseases are transmitted from one plant to another by insects, usually aphids or leafhoppers. Viruses are also very serious problems in plants that are propagated by bulbs, roots, and cuttings because the virus is easily carried along in the propagating material.

Nematodes are small eel-shaped worms that reproduce by eggs. The number of eggs produced by one female nematode and the number of generations in a season depends largely on soil temperature. Therefore, nematodes are usually more of a problem in warmer areas of the country. Most nematodes feed on the roots and lower stem of plants, but a few attack the leaves and flowers.

The life cycles of all disease organisms are greatly influenced by environmental conditions. Temperature and moisture are probably the most important factors that affect the severity of plant diseases. They not only influence the activities of the disease organism, but also affect how easily a plant becomes diseased and the way the disease develops.

How Are Plant Diseases Identified?

The extremely large number of plant diseases (about 50,000 in the United States) makes it impossible for any one person to be familiar with all of them. However, there are certain facts that help in the identification of plant diseases. The most important is to know the name of the plant that is affected.

One can then check known plant diseases to see if a similar condition has ever been reported on that plant. How the disease developed is also important. If it occurred in a very short time (overnight), it is probably not a parasitic disease, but is more likely due to some unfavourable environmental condition or chemical.

How is the condition distributed in the planting? Is it general over the area or only in one or two spots? Parasitic diseases usually do not affect a large percentage of the plants in the early stages, but start in one area and gradually spread to the other plants. And, parasitic diseases usually do not affect several different kinds of plants in one area at one time, even though some disease organisms can attack many different plants.

How Do We Recognize Plant Diseases?

The fact that diseased plants are in some way different from normal plants indicates that there are ways of recognizing plant diseases. Symptoms are outward expressions of plant diseases and include three general types: overdevelopment of tissue, such as galls; underdevelopment of tissue, such as yellowing or dwarfing; and death of tissue, such as blights and cankers.

How Are Plant Diseases Controlled?

The key to controlling parasitic diseases is to introduce an obstacle into the life cycle of the disease organism. This obstacle may be the use of disease-resistant plants on which the disease organism cannot grow. It may be to create environmental conditions that are unfavourable for disease development, or it may involve placing a protective chemical over the surface of the plant.

PLANT PATHOLOGY

DISEASE TRIANGLE

Three critical factors or conditions must exist for disease to occur: a susceptible Host plant, a pathogen, and the right mix of environmental conditions. The relationship of these factors is called the disease triangle. If only a part of the triangle exists, disease will not occur. Understanding the disease triangle helps us understand why most plants are not affected by the many thousands of diseases that exist.

PATHOGEN

Most pathogens are host-specific to a particular plant species, genus or family. For instance, blackspot of rose will not attack marigolds or lettuce. Some diseases, such as the powdery mildews, produce similar symptoms on different plants. However, the fungi involved are usually host-specific. The rose powdery mildew fungus will not infect zinnias or turfgrass or vice-versa.

SUSCEPTIBLE HOST

A susceptible host has a genetic makeup that permits the development of a particular disease. The genetic defence against a disease is called disease resistance. This resistance can be physical characteristics of the plant (fuzzy or waxy leaf surfaces), chemical characteristics (enzymes that kill pathogens and lack of enzymes) and growth patterns (ability to block off diseased tissue or outgrow damage). Plants also may be disease-tolerant. Even though infected with a disease, they can grow and produce a good crop or maintain an acceptable appearance.

The plant outgrows the disease and symptoms are not apparent or at a damaging level. It is important to remember that plants labeled as disease-resistant are resistant only to a particular disease. They are not resistant to all diseases. Resistance does not mean immunity. Under extreme circumstances, resistant plants may be infected by the disease to which they have resistance.

For disease to occur, the host plant must be at a stage of development that allows it to be susceptible to infection. For example, damping-off only affects seedlings. Botrytis is primarily a disease of buds, although it also can occur on flowers and leaves. Also, it is important that the pathogen be in a proper stage of its development to infect host plants.

ENVIRONMENTAL CONDITIONS

Certain environmental conditions must exist for disease pathogens to cause infection. The specific conditions vary for different pathogens. High moisture and specific temperature ranges, for example, are necessary for many fungal diseases. These conditions must continue for a critical period of time while the pathogen is in contact with the host for infection to occur.

Moisture, temperature, wind, sunlight, nutrition and soil quality affect plant growth. If one of these factors is out of balance for the culture of a specific plant, that plant may have a greater tendency to become diseased. For example, lilacs growing in shade are more likely to be infected with powdery mildew than those growing in full sunlight. Environmental conditions also affect the growth and spread of disease pathogens. Very dry or wet weather will have an accompanying set of diseases that thrive under these conditions.

Moisture

Moisture in the plant environment can include humidity, dew, rainfall or water from irrigation. Moisture is critical to the spread of most plant diseases. Familiar diseases, such as black spot, fireblight and apple scab require moisture to spread to and infect new host plants. Constantly wet foliage from overhead watering is a condition that promotes disease development. Seedlings grown indoors in soggy, unsterilized potting medium and pots are more prone to damping-off, a fungal disease.

Temperature

Each disease pathogen has a specific temperature range for growth and activity. There are warm-weather and cool-weather diseases. Many powdery mildew diseases are late summer, warmer temperature diseases. Temperature affects how rapidly pathogens multiply. Soil temperature can also be critical for disease infection. Cool, wet soils promote fungal root diseases. Temperature extremes can cause stress in host plants, increasing susceptibility.

Wind and Sun

The combination of wind and sun affects how quickly plant surfaces dry. Faster drying generally reduces the opportunity for infection. Wind can spread pathogens from one area to another, even many miles. Wind and rain together can be a deadly combination. Windblown rain can spread spores from infected plant tissue, blowing these pathogens to new host plants.

Sunlight is very important to plant health. Plants that do not receive the right amount of sunlight to meet their cultural requirements become stressed. This may make them more susceptible to infection.

Soil and Fertility

Soil type can affect plant growth and also development of some pathogens. Light sandy soil low in organic matter favours growth of many types of nematodes. Damping-off disease increases in heavy, cold, water-logged soils. Soil pH affects pathogen development in some diseases. Clubroot of cabbage occurs in soils with a low pH, for example. High soil pH is a factor in the development of scab on potatoes.

Fertility affects a plant's growth rate and ability to defend against disease. Excessive nitrogen fertilization can increase susceptibility to pathogen attack. It causes formation of SUCCULENT tissue and delays maturity. This can contribute to certain patch diseases in lawns. Nitrogen deficiency results in limited growth and plant stress which may cause greater disease susceptibility.

DISEASE CYCLE

Often gardeners believe that their plants have become diseased overnight. This may be true in the case of damping-off. More often, however, much has occurred before symptoms are seen. There are five stages in disease development: inoculation, incubation, penetration, infection and symptoms. The pathogen must be introduced (inoculated) to the host plant. Most pathogens cannot move on their own, but must be carried to the host plant. This is done by rain, wind, insects, birds and people. Splashing rain carries spores of apple scab fungus from infected apple leaves to uninfected leaves. Wind blows fungal spores from plant to plant. The spotted cucumber beetle transmits bacterial wilt of cucumbers when feeding.

Working in the garden when plants are wet is a common way to spread disease. Disinfesting tools requires a 9-to-1 solution of water and bleach and takes a minimum of ten minutes. Smokers can transmit tobacco mosaic virus from a cigarette to tomato plants. Seeds or cuttings from infected plants will also transmit disease. Certified seed guarantees that at the time of sale the seeds are free of all diseases. Seeds are often coated with a fungicide to prevent the transmission of surface fungal diseases.Disease-free stock guarantees that the plant is not infected with disease. This is particularly important with perennial plants, such as roses, raspberries and other small fruits.

Incubation

The second stage of disease development is incubation. The pathogen changes or grows into a form that can enter the new host plant. In many fungal diseases, the pathogen arrives on the plant as a spore which must germinate before it can grow into the plant.

Penetration

The third stage is penetration or the point at which the pathogen actually enters the host plant. Once the fungal spore germinates, it sends out thread-like tubes call hyphae. These penetrate the plant through wounds or natural pores. Wounding roots of bedding plants during transplanting provides entry for root-rotting fungi. The mouthparts of an insect also result in openings for penetration.

Infection

The fourth stage is infection. The pathogen grows within the plant and begins damaging the plant tissue.

Symptoms

As the pathogen consumes nutrients, the plant reacts by showing symptoms. Symptoms are evidence of the pathogens causing damage to the plant. Symptoms include mottling, dwarfing, distortion, discolouration, wilting, and shriveling of any plant part.

PATHOGEN SURVIVAL

Many pathogens can survive without a susceptible host even under the most unfavourable conditions. Many plant diseases survive from one growing season to the next on plant debris, seeds, alternate hosts or in soil. Because of pathogen survival, it is important to remove and properly dispose of any infected plant materials. It is also important for the gardener to know about the diseases that affect each plant throughout the home landscape, as well as the conditions needed for proper culture.

PLANT PATHOGENS

Pathogens are grouped taxonomically into related groups. It is helpful to know the type of pathogen in order to prescribe a control. Plant disease names are often very colourful and descriptive, as in fireblight, black knot and clubroot. Many disease names indicate something about the appearance of the symptoms.

Fungi

Fungi (plural of fungus) are the largest group of plant pathogens. Fungi often develop into colonies, like bread mold. Some fungi develop into large structures, such as mushrooms. Fungi spread by spores carried by wind, water or animals. They enter host plants directly or through natural openings or wounds.

They cause damage by producing substances that change or destroy plant tissues. Common fungal diseases include powdery mildew, rust, leaf spot, blight, root and crown rots, damping-off, smut, anthracnose, and vascular wilts.

Powdery Mildews

These fungal diseases are very HOST-SPECIFIC. It is common to see powdery mildew on lilacs, roses, zinnias, melons, beans, cucumbers, turfgrass and many other plants. The symptom is a white or gray, powdery growth on leaves and stems. Powdery mildew will not usually kill a plant; however, it may weaken plants, and it is unsightly. Providing adequate sunlight and air circulation to lower relative humidity and using fungicide sprays for high-profile plants help control powdery mildew.

Rust

Rusts are also host-specific. They produce masses of easily noticed, orange or dark red spore masses on leaf tissue. Rusts are commonly seen in our region on turfgrass, hollyhocks, snapdragons, hawthorns and the small fruits. Rusts usually develop during cool weather. They are spread by wind and splashing water. Rusts require water to reproduce and infect host plants. Controls include watering early in the day and applying protective fungicidal sprays.

Leaf Spot and Blight

Rose growers are very familiar with black spot, a common leaf-spotting disease of roses. Scab on crabapple and pyracantha are leaf-spotting diseases that can cause defoliation.

Botrytis flower blight occurs on annuals and some perennials during wet seasons. Anthracnose of shade trees is a blight frequently seen on sycamores, maples, ashes and oaks. Fungi-causing tree diseases are spread by wind and splashing water.

Grow resistant cultivars, maintain plant vigour, enhance leaf drying by proper pruning and water early in the day to help avoid leaf spotting and blights. Fungicides provide protection from infection when applied before infection by fungal spores.

Root and Crown Rots

These diseases, which include water mold diseases, attack a wide variety of plants. They cause damage to roots and crowns of plants.

Damage to some of the root system results in poor growth, yellowing or stunting of the plant. Root rot fungal pathogens are found in almost all soils. However, they do not survive as well in well-drained soils. These organisms can live in soil for years in a dormant state. Proper cultural practices, such as correct planting depth and improving drainage, are important controls.

Stem and Twig Cankers

Cankers may be compared to sores on a stem or twig. The tissue may be raised or sunken, discoloured or split open. Twig cankers on junipers and leaf and stem blight of pachysandra are common examples. Most of the canker diseases are not controllable with fungicides.

Maintaining plant vigour, protecting from injury, removing diseased parts and growing plants suitable for the site help prevent canker problems.

Vascular Wilts

Dutch elm disease is the classic wilt disease. The fungus invades the vascular system of the plant and clogs it, stopping the flow of water and nutrients. Verticillium wilt is common on trees, shrubs, vegetables and fruits. Discolouration of the vascular tissue sometimes can be seen when you cut diagonally through a branch. There are no effective chemical controls for wilts. Growing resistant species, maintaining plant vigour and sanitation provide the only controls.

Smuts and Molds

Smut fungi produce large amounts of black sooty spores which show up as unsightly dark brown or black blotches on leaves, stems and fruit.

Bacteria

Bacteria are microscopic, single-celled organisms that multiply rapidly. They produce chemicals that destroy plant tissue or cause abnormal growth. Bacteria enter through wounds, stomata or other natural openings.

Bacteria can cause leaf spotting as on English ivy. Fireblight on crabapples, pears, mountain ash, hawthorn, cotoneaster and pyracantha is a bacterial disease. Bacteria also cause diseases, such as crown gall on stone fruits and euonymus and wilts in geraniums and cucurbits (cucumbers, squash and melons).

Disease-free stock and resistant cultivars plus sanitation help prevent bacterial diseases. When cutting out bacterial infections, sterilize pruners with 70% alcohol between cuts. Streptomycin and fixed copper sprays may help slow the spread and development of bacterial diseases. However, chemical controls for bacterial diseases are less reliable than for many fungal diseases.

Mycoplasmas

Mycoplasmas are disease agents that were not accurately identified until the 1960's. Previously many mycoplasma-caused diseases were thought to be caused by viruses. A group of diseases caused by mycoplasmas is called YELLOWS diseases. Yellows disease is usually spread from one host plant to another by an insect vector. Leafhoppers are common vectors for mycoplasmas.

Viruses

Viruses are organisms so small that they can only be seen under an electron microscope, magnified 2,000 to 3,000 times. Viruses multiply only within living cells of host plants. Viruses are spread by insects, nematodes and humans. Symptoms include vein banding.

Mosaic, flecking or spotting on foliage and abnormal growth, similar to herbicide damage. Tobacco mosaic virus on tomatoes and leaf mosaic on dahlias are common viral diseases. Control is limited to removing and destroying infected plants.

Nematodes

Nematodes are microscopic, worm-like organisms. They most commonly feed on plant roots, but some nematodes invade leaf tissue. Nematodes suck out liquid nutrients and inject damaging materials into plants.

They injure plant cells or change normal plant growth processes. Symptoms of nematodes include swelling of stems or roots, irregular branching, deformed leaves, lack of blossoming and galls on roots. Nematodes can facilitate the entry of viruses and fungi into plants. Control efforts are difficult and provide limited success. Chemical controls must be done by professional pesticide applicators. A specialized soil test for nematodes indicates the type of nematode.

DIAGNOSIS OF PLANT DISEASES

To correctly diagnose plant disease problems, follow a few basic steps. View the plant and its environment from various perspectives. The most obvious place to look first is up close. Use a hand lens if necessary. Don't stop at the first or most obvious symptom; check for more.

You want to find all of the symptoms. Look for symptoms on leaves, stems, roots, flowers and fruits. Cut open a branch or stem to look for vascular problems. Vascular problems show as discolouration of vascular tissue, leaf or stem wilting and sudden wilting of a section or a total plant.

General View

Stand back and look at the overall picture. Consider the total environment: weather, soil, stage of development for plant and pathogens, cultural practices and condition of other plants in the area. A plant growing in the wrong location may be stressed. Consider pesticide applications, recent construction or digging, and weather conditions.

Time

Determine when the symptoms became apparent. The onset of a problem may be due to a cultural practice, the seasonal appearance of a disease or insect, or a weather-related event. Remember that long-term stress is slow to appear, taking a year or more at times. Is the problem spreading? This may indicate it is a pathogen. Are plants of other species affected? Diseases are usually species-specific. Problems caused by environmental factors do not spread, although the symptoms may become more severe.

Knowledge and Experience

You must know what the plant should look like to be able to determine abnormalities. Check the references to see what problems are typical for a particular species. Gather all the information you can to help you make the diagnosis.

Remember that there is usually no single cause. There may be a primary cause; however, it may be associated with cultural or environmental conditions. Just as there is probably no single cause, there is usually no single symptom. Search for all of the symptoms. Orderly thinking and good questions are the key to accurate diagnoses. When in doubt about a diagnosis, turn to agents, state specialists or the Pest and Plant Disease Clinic for assistance or a second opinion.

Nonliving or Abiotic Agents

Nonliving or abiotic agents can indirectly result in plant problems. Additionally, several factors in the plant's environment can produce disease-like symptoms: weather extremes, high winds, high or low temperatures, nutrient deficiencies, physical damage and poor cultural conditions. Frost often damages buds and leaves in early spring. Hail can cause leaf spotting or holes. Drought and high winds result in wilting and in extreme cases, browning and curling. Air, water and soil pollution affect plant health and can produce disease-like symptoms. Soil imbalance, resulting from construction or other dumping, or misapplied garden chemicals can cause damage and disease-like symptoms.

Dumping of household, automotive and industrial chemicals can also produce plant damage. Plant disease can result from a combination of abiotic agents and biotic agents. Plants may be initially placed under stress by nonliving agents. This creates a susceptibility in plants for attack by living agents. Drought may damage roots which then are more likely to be infected by fungal diseases.

Random or Uniform Patterns

Random distribution of symptoms on injured plants is usually caused by a biotic factor, such as infectious disease pathogens or an insect/animal. Uniform patterns are generally associated with abiotic or noninfectious agents like pesticides, fertilizers, environmental or site stress and mechanical damage.

Collecting Specimens

Plant specimens that are to be diagnosed should be taken from the area where symptoms are showing on living tissue. Dead plants are often invaded by secondary pathogens which may hide the original problem. Collect several representative samples showing various stages of disease development. A generous sampling will assist in diagnosis. If possible, collect the entire plant, including roots. Wrap the specimens in dry paper.

Do not moisten them or seal them in plastic wrap or plastic bags. Never mix different specimens in a single bag. A fresh sample is required. Complete the diagnostic form as thoroughly as possible. This will result in better diagnosis.

DISEASE CONTROL

By the time disease symptoms appear, disease pathogens are inside the plant and usually beyond control. Therefore, it is important to prevent penetration of pathogens. Consider the following methods of disease control.

Avoid

Avoid certain diseases through choice of appropriate site and planting time, purchase of disease-free stock and cultural practices that do not favour disease infection. Insect control can be important for controlling spread of viruses, mycoplasmas, bacteria and fungi. Mulching can help control diseases by preventing the contact of foliage with soil.

Exclude

Exclude insect vectors with row covers. Don't bring diseased plants into the garden.

Use Resistant Or Tolerant Plants

Appropriate selection of species or cultivars will avoid or minimize problems. Remove all diseased plants, including alternate hosts. Immediate removal of diseased plants or plant parts reduces the chance of the disease spreading.

Selective pruning and careful sanitizing of pruning equipment can prevent spread of disease. Rotate crops to avoid soil-borne diseases. Use best cultural practices. Follow recommended cultural practices.

TREAT

Treat plants with recommended chemicals to prevent infection. Some treatments will stop the spread of the pathogen.

Chemical Treatment of Plant Disease

To use fungicides and bactericides effectively, follow these 4 steps:

- Be sure that the diagnosis is correct.
- Choose a recommended control. Remember, a fungicide controls fungal diseases and bactericides only control bacteria. Read the product label and follow the instructions precisely. Make sure the plant you will be treating is listed on the label.
- Apply the control correctly following all label instructions To be effective, fungicidal sprays must be applied properly. They work by creating a barrier that pathogens cannot penetrate. Good coverage is important. Give special attention to the underside of leaves, especially lower leaves. The barrier's effectiveness is affected by how well the spray spreads and sticks. Many fungicides have Spreader-Stickers added to the product. If you are not sure about coverage, observe the pattern spray deposits.
- Apply at the recommended time. In many cases, fungicides are a protectant and must be applied just before the pathogen is able to infect the plant. As the plant grows, additional sprays may be necessary to protect the new growth.

Fungicides are developed so that they degrade fairly rapidly. They are affected by rain, sun, and microbial activity. Excessive rainfall or growth will require shorter intervals between spraying. New shoots should be treated. Product label guidelines usually indicate a range of 7 to 14 days.

Know exactly what disease pathogen or combination of pathogens and abiotic agents are causing a plant problem. Fungicides and bactericides are effective against specific diseases under specific conditions. In the past, general fungicides were used. Many pathogens developed resistance to these general pesticides.

Use common sense in analyzing a situation. If an expensive or sentimental plant is experiencing problems, seek the help of professionals at the Pest and Plant Diagnostic Clinic.

PEST-RESISTANT CROPS

One of the mainstays of integrated pest management is the use of crop varieties that are resistant or tolerant to insect pests and diseases. A resistant variety may be less preferred by the insect pest, adversely affect its normal development and survival, or the plant may tolerate the damage without an economic loss in yield or quality.

Disease-resistant vegetables are widely used, whereas insect-resistant varieties are less common but nonetheless important.

Examples include varieties of wheat which have tough stems that prevent development of the Hessian fly and cucurbits (squash, cucumbers, melons) that have lower concentrations of feeding stimulants (cucurbitacins) for cucumber beetles.

Commercial corn lines with increased concentration of a chemical called Dimboa display resistance to European corn borer, a major pest throughout the world. In the case of cabbage, the only reliable method of controlling onion thrips is through the use of resistant varieties. Advantages of this tactic include ease of use, compatibility with other integrated pest management tactics, low cost, and cumulative impact on the pest (each subsequent generation of the pest is further reduced) with minimal environmental impact.

The development of resistant or pest-tolerant plant varieties, however, may require considerable time and money, and resistance is not necessarily permanent. Just as insect populations have developed resistance to insecticides, populations of insects have developed that are now able to damage plant varieties that were previously resistant.

CHEMICAL CONTROL

If all other integrated pest management tactics are unable to keep an insect pest population below an economic threshold, then use of an insecticide to control the pest and prevent economic loss is justified. In most cropping systems, insecticides are still the principal means of controlling pests once the economic threshold has been reached.

They can be relatively cheap and are easy to apply, fast-acting, and in most instances can be relied on to control the pest(s). Because insecticides can be formulated as liquids, powders, aerosols, dusts, granules, baits, and slow-release forms, they are very versatile.

Insecticides are classified in several ways, and it is important to be familiar with these classifications so that the choice of an insecticide is based on more than simply how well it controls the pest. When classified by method of application (site of encounter by insect), insecticides are referred to as stomach poisons (those that must be ingested), contact poisons, or fumigants. The most precise method of classifying insecticides is by their active ingredient (toxicant).

According to this method the major classes of insecticides are the organophosphates, chlorinated hydrocarbons, carbamates, and pyrethroids. Others in this classification system include the biologicals (or microbials), botanicals, oils, and fumigants. Insecticides may be divided into two broad categories: (a) conventional or chemical and (b) biorational. In this guide we define conventional or chemical insecticides as those having a broad spectrum of activity and being more detrimental to natural enemies. In contrast, insecticides that are more selective because they are most effective against insects with certain feeding habits, at certain life stages, or within certain taxonomic groups, are referred to as "biorational" pesticides. These are also known as "least toxic" pesticides.

Because the biorationals are generally less toxic and more selective, they are generally less harmful to natural enemies and the environment. Biorational insecticides include the microbial-based insecticides such as the Bacillus thuringiensis products, chemicals such as pheromones that modify insect behaviour, insect growth regulators, and insecticidal soaps. The majority of insecticides fall into the chemical category because they are typically more effective and can usually be used to control several pests. This provides the economic justification needed for the research and development of such products. In contrast, the more specialized market of the biorationals makes their long-term economic return less favourable.

Despite the advantages of conventional insecticides, the problems associated with their use have been well documented. These include the resurgence of pest populations after decimation of the natural enemies, development of insecticide-resistant populations, and negative impacts on nontarget organisms within and outside the crop system. One of the more serious problems is the development of resistance. Many insect pest species now possess resistance to some or several types of insecticides, and few chemical control options exist for these pests.

If all other integrated pest management tactics are unable to keep an insect pest population below an economic threshold, then use of an insecticide to control the pest and prevent economic loss is justified. In most cropping systems, insecticides are still the principal means of controlling pests once the economic threshold has been reached. They can be relatively cheap and are easy to apply, fast-acting, and in most instances can be relied on to control the pest(s). Because insecticides can be formulated as liquids, powders, aerosols, dusts, granules, baits, and slow-release forms, they are very versatile.

To compare pesticide use patterns, the dose, formulation or per cent active ingredient of the product, and the frequency of application must also be considered and are used to calculate the EIQ Field Use Rating. Pesticides with a lower EIQ Field Use Rating have a lower environmental impact. The EIQ concept and techniques are evolving, but this, or a smilar rating, will provide information for appropriate pest management decisions in the future. Natural enemies are generally more adversely affected by chemical insecticides than the target pest. Because predators and parasitoids must search for their prey, they generally are very mobile and spend a considerable amount of time moving across plant tissue.

This increases the likelihood that they will contact the insecticide. When an insecticide is applied, ideally only the target pest(s) should be affected. The goal is to maximize pest mortality while minimizing harm to natural enemies.

The following factors, some of which are used when determining the EIQ, influence the impact of insecticide applications on natural enemies:

- Spectrum of activity. Insecticides toxic to most insects (broad-spectrum) will more adversely affect natural enemies than materials that are narrow-spectrum or selective for specific insect species or life stages. Most insecticides in use today have a broad spectrum of activity.
- Residual (half-life) activity of insecticide. Insecticides that remain toxic to pests for a long time and remain on the treated surface will have a similar effect on natural enemies.
- Coverage and formulation of insecticide. Full coverage sprays will generally have a greater impact on natural enemies than directed sprays, systemics, or bait formulations. Spot or edge treatments directed at localized pest infestations or to a specific plant surface most often occupied by the pest will be less detrimental than those applied to the entire field or plant.
- Dosage and frequency of application. Higher rates and repeated applications will have a more detrimental impact on natural enemies.
- Timing of application. Applying insecticides when natural enemies are not abundant or are less susceptible, such as when immatures are encased in host eggs, can be helpful.
- Susceptibility of natural enemy to insecticide. Some natural enemies are inherently more resistant to insecticides than others, and some populations of natural enemies have been selected either naturally or through the efforts of researchers to possess higher levels of resistance. For example, strains of predaceous mites, rove beetles, lacewings, and species of parasitoids have been selected for increased levels of resistance to insecticides, thereby increasing the likelihood that some will survive insecticide treatments directed at pests. Some of these pesticide-tolerant strains are commercially available. Applications of some fungicides directed at

controlling plant diseases can also reduce the incidence of fungal diseases of insects. In general, little can be done about this incompatibility except to use the fungicide only as needed.

In addition to killing natural enemies directly, insecticides may also have sublethal effects on insect behaviour, reproductive capabilities, egg hatch, rate of development, feeding rate, and life span. Fungicides and herbicides may also have lethal and sublethal effects on natural enemies.

CULTURAL CONTROL

There are many agricultural practices that make the environment less favourable to insect pests. Examples include cultivation of alternate hosts (e.g., weeds), crop rotation, selection of planting sites, trap crops, and adjusting the timing of planting or harvest. Crop rotation, for example, is highly recommended for management of Colorado potato beetle. The beetles overwinter in or near potato fields and they require potato or related plants for food when they emerge in the spring.

With cool temperatures and no suitable food, the beetles will only crawl and be unable to fly. Planting potatoes well away from the previous year's crop prevents access to needed food and the beetles will starve. The severity and incidence of many plant diseases can also be minimized by crop rotation, and selection of the planting site may affect the severity of insect infestations.

Trap crops are planted to attract and hold pest insects where they can be managed more efficiently and prevent or reduce their movement onto valuable crops. Early planted potatoes can act as a trap crop for Colorado potato beetles emerging in the spring. Since the early potatoes are the only food source available, the beetles will congregate on these plants where they can more easily be controlled.

Adjusting the timing of planting or harvest is another cultural control technique. The earlier planted processing tomatoes grown in the western United States are far less likely to be infested by the tomato fruitworm than those planted later in the season. It is also important to use pest-free transplants. Some vegetable crop transplants can be infested with insect pests, and growers using these transplants are put at a considerable disadvantage.

PHYSICAL AND MECHANICAL CONTROL

The use of physical barriers such as row covers or trenches prevents insects from reaching the crop. Row covers can help prevent early-season damage to cucurbits by cucumber beetles, and plastic-lined trenches are effective in trapping large numbers of dispersing Colorado potato beetles in the spring and fall. Cold storage is also considered a physical control and, although it does not necessarily kill the insect pests, it at least stops their development and further feeding on the stored crop. Other methods include hand picking of pests, sticky boards or tapes for control of flying insects in greenhouses and various trapping techniques.

Chapter 2

Pathogens Associated to the Seeds

As an introduction we will remind some aspects of the morphology and anatomy of the seed, in order to understand in a better way the effect pathogens have in the seed and later, the plant originated from it.

THE SEED

The seed consists of three basic parts:

- Embryo
- Storage tissues
- Seed coat.

The embryo has a reproductive function, being capable of initiating cellular divisions and growth. It is the most important part of the seed. It is an axis that originates growth in two directions, with the objective of originating a root and a shoot. Usually, the embryo is very small, compared to other parts of the seed. It is like a miniature plant. It is consists of a radicle, plumule, one or two cotyledons, and hypocotyl or epicotyl, depending on the type of plant.

Energy storage is localized in the cotyledons, in the endosperm or the perisperm. Cotyledons originate from the zygote and are part of the embryo. In many species, storage substances are localized in the cotyledons, and the embryo develops absorbing all the endosperm. Episperm is the seed cover, consists of two layers, the testa and the endopleura. The outer layer is the testa; it can be stony, leathery, membranous or fleshy.

Over the testa we can recognize: the hilum, scar or point of attachment of the seed to the funiculus, water penetrates easily through it; Micropyle, the point upon the seed at which was the orifice of the ovule through which the pollen tube enters; raphe, suture originated by the part of the funiculus that is fused along the side of the ovule. The endopleura is the inner layer; it is thin and generally whitish. Teguments, testa or protective covers delimit the seed. They are formed by one or more layers of cells originated from the ovule integuments and sometimes from the pericarp made from the walls of the ovary.

Regarding the healthiness or pathology of the seed we will consider three aspects that allow us to understand processes associated to seed-borne diseases: infection of the seed, infection of the plant and steps that can be taken to reduce the damage caused by this relationship.

Seed Infection Mechanisms

The area of science that studies the relationship between pathogens and seeds is Seed Pathology. It does not only identify the pathogens, it also includes the role of the seed as source of inoculum, the survival of the pathogen and the actions taken to control the pathogens associated to it. It uses the knowledge of General Pathology, Microbiology and Seed Analysis. To obtain a perspective of seed borne diseases, seed borne microorganisms can be considered fewer than four classes.

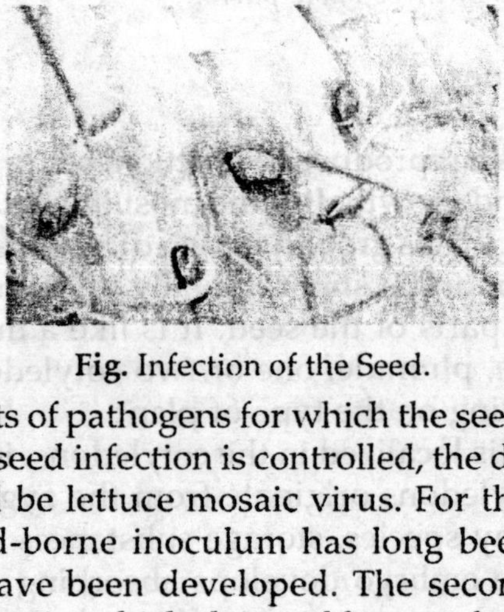

Fig. Infection of the Seed.

The first consists of pathogens for which the seed is the main source of inoculum; when seed infection is controlled, the disease is controlled. An example would be lettuce mosaic virus. For these pathogens, the importance of seed-borne inoculum has long been recognized, and control practices have been developed. The second class consists of important pathogens in which the seed borne phase of the disease is of minor significance as a source of inoculum.

Examples are those in which the crop residues in the field were the major source of inoculum. The third and largest group of seed borne microorganisms consists of those that have never been shown to cause disease as a result of their presence on seeds. The fourth class is a group of microorganisms that can infect the seed either in the field or in storage and reduce yield and seed quality. Examples of field fungi are *Diplodia, Fusarium, Cladosporium*, etc.

The storage fungi *Aspergillus* and *Penicillium* can invade most types of seeds under high-moisture storage conditions. The process of seed infection is influenced by the conditions under which the crop grows.

Between the facts that influence in the process of infection are: the host and its genotype, the pathogen and its genotype end environmental facts. An infected seed will not always an infected seed will be the cause of the infection of the infection in the plant which originates, so, it is to say that the infection caused from an infected seed is an exception, not something to happen generally. There are two situations, Systemic infection of the seed and Contamination or Infestation of the seed.

Systemic Infection of the Seed

The establishment of a pathogen in any part of the seed is refereed as seed infection. It can be systemic, by the vascular system or plasmodesmata or directly by natural or artificial wounds. The same pathogen can infect the seed using one or more of these mechanisms.

For example: *Xanthomonas campestris* pv. *phaseoli* can infect seed through the vascular system, by natural openings (From the pod suture goes to the funiculus, then to the raphe and tegument, or it can also happen thought the micropyle).

Systemic Infection through Flowers, Fruits or Funiculus

Most of the systemic seed-borne bacteria and fungus reach and infect the embryo through the flower or from the peduncule of the fruit, via funiculus. Viruses go to the embryo from the systemically infected mother plant and the infected or contaminated pollen. They rarely reach the embryo during the formation of the seed or the embryo itself.

Examples of some infections that occur through the vascular system are: Some forma specialis of Fusarium oxysporum in pumpkin, pea and tomato; Plasmopara halstedii in sunflower; Septoria glycines in soy; Verticilium dahliae in spinach and sugar beet; Certain pathovares of Xanthomonas campestris in bean, cabbage, rice and sweet pepper; and Pseudomonas syringae pv. Lachymans in cucumber.

Penetration through the Stigma

During the infection, some pathogens follow the same path as the pollen grains do. Spores of some fungus reach the stigma and germinate, producing an hypha that reaches the ovary through the style, where they can stay as a dormant mycelium until seed germination.

For example: Ustilago nuda and U. triticci in barley and wheat, and Alternaria alternata in sweet pepper. Viruses can infect through infected pollen, the male gamete carries the virus, when joining the ovule it generates an infected embryo. If both, the male and female gamete are infected they can even produce an infected endosperm.

Penetration through the Wall of the Ovary

Some fungi, like *Ustilago nuda* and *U. triticci* penetrate through the wall of the ovary as a result of the germination of the Teliospores on the stigma or the wall of the ovary. The pro-mycelium goes through the wall and other tissues until it reaches the embryo. In some other cases, penetration occurs through breakages on the testa, establishing itself in the endopleura or the endosperm. In fleshy fruits, like cucumber, melon, eggplant, tomato, sweet pepper and others, contamination can occur directly through the funiculus or in the tegument, during the process of seed formation. Examples of this are Colletotrichum lagenarium in watermelon and Rhizoctonia solani, when it invades fleshy fruits, it is capable of infecting from the placenta and penetrate to the developing ovule or seeds that are still in its formation process and have not lignified its cover.

Penetration through wounds and Natural Openings

Natural openings like the hilum and the micropyle or wounds generated during the thresh are spots where pathogens like *Xanthomonas campestris* pv. *phaseolicola* in bean and *Pseudomonas syringae* pv. *lachrymans* in cucumber.

SEED CONTAMINATION OR INFESTATION

Contamination or infestation refers to the passive relationship of a pathogens and seeds. The pathogen itself or parts of it can stick or can get mixed with the seeds during any of the processes during seed recollection: harvesting, extraction, thresh, selection and packing.

Pathogens that Stick to the Surface of the Seed

Pathogens that stick to seeds during harvest or postharvest do so by their spores (Clamidospores, Oospores, Teliospores, Uredospores), bacterial cells and in some cases, virions. Spores of the following fungi can be carried on seed coat surfaces: Alternaria brassicae and A. brassicicola in crucifers; A. longipes in tobacco; A. radicina in carrot; Ascochyta pinodella in pea; Drechslera sorokiniana and D. oryzae in rice; D. avenae in oats; Fusarium oxysporum f. sp. callistephi in China aster; Sclerotia of Rhizoctonia solani in eggplant, pepper, and tomato; Tilletia caries, T. foetida, T. contraversa, and Urocystis agropyri in wheat. A number of bacteria, such as the following, contaminate seed surfaces: Corynebacterium flaccumfacjens pv. flaccumfacjens in bean,

Pseudomonas syringae pv. phaseolicola in bean, P. syringae pv. lachrymans in cucumber, P. syringae pv. tomato in tomato, C. rnichiganense pv. michiganense in pepper, *Xanthomonas campestris* pv. *campestris* in cabbage Also some viruses as Tobacco mosaic virus, Tomato mosaic virus, Pepper Mosaic virus.

ACCOMPANYING CONTAMINATION

This type of contamination refers to physical mixing of the seed with pathogen's propagation organs like the sclerotium, nematode's galls, contaminated plant parts or soil particles containing pathogens.

Structures of the Pathogens

Some fungi produce resistance structures named sclerotiums that are made of compacted mycelium. Under certain conditions of temperature and humidity it can germinate, it is common that this occurs together with the seed, producing its infection, by direct penetration or by the spores generated by the fructification organs (Apothecium or Perithecium). These spores are carried by the wind and taken to susceptible tissues of the plant, generally, some parts of the flower. Sclerotiums have many shapes and during thresh can be easily included with the rest of the seed. Some examples of this are Sclerotinia sclerotiorum in horticultural, grain and flower crops, Sclerotinia cepivorum in garlic and Claviceps purpurea in barley.

Mix with Infected Plant Parts

Infected plant's parts and residues can carry fructifications or spores of fungi and bacteria. This situation is particularly common in cereal, forage and oily crop's seeds. Examples of this are Septoria nodorum in wheat, Puccinia malvacearum in hollyhock (malvarrosa), Colletotrichum trifolii in clover, Erwinia carotovora pv. carotovora in tobacco, Pseudomonas syringae pv. phaseolicola in bean and Sclerotinia sclerotiorum in sunflower, soy and peanut.

Soil

Seeds can be mixed with contaminated soil and, carry micro-sclerotiums of *Macrophomina phaseolina* in dirty seeds of kidney bean, Verticillium oxysporum f.sp. phaseoli in cotton seeds, Plasmodiophora brassicae in turnip and Fusarium oxysporum f. sp. phaseoli in bean.

SEED TRANSMITION

We saw how the seed gets infected or infested. Now we will see how these seeds can produce an ill plant. This is to say, how the inoculum goes from the seed to the plant. We call seed-borne pathogens only those that can produce an infection. Because they have to be distinguished from those that can associate to the seed but do not produce an infection, these ones are called pathogens not transmitted by seed.

The virus that causes curly top in sugar beet can be present in the perisperm but does not cause an infection in the plant. In general terms, infection can be classified into systemic and non systemic. It is systemic when the pathogen introduces itself to the plant when the seed germinates, and develops with it. Non systemic infection occurs when there is a localized infection caused by the pathogen in the seedling at the stage of pre or post emergency, in this situation there are no systemic symptoms.

Systemic Transmition

This type of infection can be produced by pathogens that are carried with the seed in various parts, like the embryo, endosperm or episperm, or by contamination of the outer portion of this one.

Infection of the Embryo

When the seed germinates, if the embryo is infected, the pathogen initiates its growth together with the plant. Symptoms can show up during different stages of development.

In the case of *Ustilago triticci*, the mycelium grows together with the plant and expresses symptoms only at the flowering stage, which are that all the tissues of the spike, exept the rachis, are replaced by spores.

Some pathovares of *Xanthomonas campestris* that infect cabbage, bean or sweet pepper move between the cells of the host until they reach the vascular system and produce symptoms in leaves or stems. Most of seed-borne viruses persist inside the embryo. Its multiplication and movement accompanies the plant during its development, and there can be symptoms at any stage, from the formation of first leaves until flowering or fructification.

Non Embryo Infection

Infection of the episperm (testa and endopleura) occasionally conducts to a systemic infection. Some bacteria, like Corynebacterium michiganense pv. michiganense in tomato and Xanthomonas campestris pv. campestris in cabbage, penetrate through stomata of cotyledons, and from there, reach the vascular system, initiating the systemic infection.

Episperm Contamination

In some few cases this type of contamination conducts to a systemic infection. Most of these exceptions are fungi that are highly specialized in their pathogenesis and produce in cereals the so called smuts, rusts or mildews.

In these cases, generally, spores are carried outside the seed, they germinate, penetrate the coleoptile, and start a systemic infection. This can occur directly o by a more complex system of haploid hypha fusion in genera like Tilletia, Ustilago, Uromyces, Sclerospora, Pernospora and Puccinia.

Non Systemic Transmition

Non systemic infection is very common, and in the same way as systemic infection, it can come from an infection, outside contamination or by pathogens mixed with the seeds.

Infection of the Embryo

This case is restricted to some pathogenic fungi that maintain itself in the embryo or the episperm as hypha inside the seed. Primary infection starts as injuries in the cotyledons or primary leaves, stems or petiole. Fungi fructifications (Pycnidium, Acervulus) can develop on these organs, under certain favorable conditions (temperature and humidity).

These fructifications produce spores that, with the action of water and wind, disperse the disease to other parts of the plant and other plants. This occurs in *Ascochyta pisi* in pea, *Colletotrichum lindemuthianum* in bean and *C. truncatum* in soy.

Infection of the Episperm

Generally, seeds in which the episperm is infected, do not geminate, or germinate and contaminate the soil. Rarely produce a systemic infection, but can infect the seedling from the outside. Over the injuries, new inoculum is produced; this one infects other parts of the plant or other plants.

Septoria nodorum in wheat under high humidity conditions forms Picnidium in the coleoptile, whose spores disseminate the disease to other plants.

Contamination of the Episperm

Contamination outside the seed's testa can produce healthy seedlings, but the inoculum infects the soil and from there it can cause infections at more developed stages of the plant. Wheat and rice grains that are contaminated with Teliospores of Neovossia indica in wheat and N. torrida in rice, produce Esporidisporas in the soil, that can be carried by the wind and infect flowers, originating the smuts.

Accompanying Contamination

The mixture of seeds with sclerotiums of the fungi or contaminated soil particles produces non systemic infections, at any stage of the plant growth and development. It is common for *Sclerotinia sclerotiorum* to produce mycelium from the sclerotiums. These mycelium can directly infect the plant or produce fructifications that produce spores that can be carried by the wind and infect flowers of different species, like sunflower, soy, peanut, etc.

Prevention of Seedborne Diseases

Prevention of seed-borne diseases is based in two main facts. On one hand there is the usage of material free from pathogens and contaminants that could originate a disease and on the other there are the treatments to eliminate the inoculum in the seed or propagule. When using material free from pathogens it is important to have adequate pathogen detection methods, this is to say, sensible and specific. No production of health quality seed can be done without valid methods to detect the possibility of contamination. Some treatments can be necessary to eliminate the pathogens from the seeds, or at leafs, maintain an economically viable balance.

Health Testing of the Seed

One of the most relevant aspects about controlled health quality seed is refereed to the detection methods used on every case. ISTA (International Seed Testing Association) appeared in 1924, by was only interested in seed's gene purity and germination aspects. But detection methods were applied.

According to the criterion of each laboratory, so their results were not comparative. So, in 1957, PDC established a comparative health testing programme and standardize applied methods. The complexity for comparison of these methods made a considerable step back on the publication of working sheets (protocols). ISTA approved methods became Rules. By 1999 only 64 protocols were being compared, and only 14 of them were accepted as Rules.

As the solutions proposed by ISTA did not satisfy the needs of international seed industry, this industry, in 1994, organized The International Seed Health Initiative for Vegetables (ISHI-Veg) chartered by the vegetable seed industries in The Netherlands and France. The

Initiative was soon joined by the seed companies in the United States, Israel and Japan. This group represents the production of over 75% of the worlds vegetable seed supply. ISHI published many protocols, organized by crop. For horticultural crops, the ISHI-VEG, the Seed Health Testing Methods

ISHI-Veg Status

- An ISHI Validated Reference Method is a method that has been through the ISTA/ISHI comparative testing process and is being published as ISTA wn ISHI Reference Method is a method that has been through ISHI comparative testing and is under review by the ISTA/ISHI reviewers for publication as ISTA Working Sheet.
- An ISHI Accepted Method is a method that is commonly used by the seed industry for determining seed health. The method has been published in a journal, as a working sheet and/or is publicly available for comparative testing and commonly in use by the seed industry. The method is in comparative testing or aspects of the test are being tested for inclusion or deletion from the method.
- An ISHI Reviewed Method is a method that is publicly available or published, documented and prioritized by the ITG, but not subjected to a comparative test at this time. The method usually is accepted by other agencies or groups.

In 1995, during the second ISTA-PDC symposium it was decided that a Joint ISTA/ISHI Guidelines for Comparative Testing of Methods for Detection of Seed Borne Pathogens would be edited.

Through this combined effort many laboratories, organized under this alignments, the ISTA developed the Manual of Seed Health Testing Methods that has two sections:

- Validated Test Methods
- Peer Reviewed Methods.

The transition process from working protocol to ruled method is long and laborious, so working protocols can be considered as adequate methods for seed analysis. Nowadays, it is common that technological advances may become obsolete even before they are published.

The protocol to analyse *Xanthomonas campesrtis* pv.*campestris*, which causes black rot in crucifers, was published in 1982, there, it was recommended one minute to extract the bacteria from the seeds to starch culture medium. 2 years later, this time was raised to 2 hours, and other specific culture mediums. In 1987 the protocol was replaced, again in 1996 and at present time again it is under revision.

Minimum requirements to Produce Certified Seed

The Organization for Economic Cooperation and Development (OECD) establishes steps to be taken in order to produce basic and certified seed. Between these steps it is emphasized that generations from the mother material to the basic material are strictly limited in number. The number of certified seed generations, after the basic seed, varies from crop to crop and technical conditions.

In both, sexually and vegetative reproduced species, the schemes include similar steps. It generally starts with selection of a plant with genetic and health advantages. If health condition was not satisfying, certain actions should be taken in order to release it from considered pathogens. This Mother Plant is maintained under strict special security conditions in order to maintain healthiness (isolated glasshouses or chambers).

In the Certification Schemes two particularities abut seed borne pathogens should be considered, those transmitted exclusively with the seed and those that can also be transmitted with contaminated soil and plant rests. Diseases transmitted exclusively with the seed can be controlled by using clean healthy seed.

This seed can be obtained by Seed Certification Programmes or by treatments that are really effective, like *in vitro* culture, thermotherapy, chemotherapy or a combined action of these, for systemic pathogens.

Tolerance levels in certified seed depend on the multiplication speed and dispersion of the inoculum of the pathogen of the considered disease. If a small amount of inoculum may be epidemic, like Anthracnose and the bean bacterial blight, if levels above zero chemical control may not be enough to stop the disease.

But pathogens whose inoculum multiplies slowly generation after generation, the disease may be evident only several generations later, so, if contamination levels are above zero may be accepted in adequate control methods are applied to the seed. This is what happens in wheat, with the Loose Smut, continuous treatments diminish inoculum levels, and make it perfectly tolerable.

New Techniques for Seed Health Analysis

Pathogen detection methods have suffered great changes since 1957. Back then it was mainly targeted to fungi, and based exclusively on incubation and identification, or the cultivation of these seeds and counting the percentage of unhealthy seedlings. These analyses were very laborious and required a high infection level in order to be detected. Nowadays, technological advances have allowed detecting even low infection levels, evaluating 1,000 to 10,000 seeds at once. Specificity of culture mediums, usage of antibiotics and other products has allowed isolating the pathogen from the seed.

Together with this, the use of reactives as mono or polyclonal antiserums and molecular markers is also available. The Lettuce Mosaic Virus (LMV), back in 1950 the symptoms were observed over many thousands of seedlings from the seed that were going to be analyzed.

Later, technical advances determined that 30,000 seedlings were needed, and even later, in 1983, an alternative to the *Chenopodium quinoa* test was discovered, the ELISA test.

Watermelon Fruit Blotch caused by bacteria is a severe problem in glasshouse winter production. It was determined that a single contaminated seed in 9,000 of them was enough too widespread the bacteria to the rest of the plants. That is the reason why screening was done over 10,000 seedlings. Nowadays PCR technique is used.

Chapter 3

Plants and Soil

WHEN a soil is sick, either because its beneficial bacteria do not perform their functions properly, or because of abnormalities in its chemical or physical properties, careful treatment and proper cultural methods may restore it to health. But when a soil becomes sick and unproductive because parasitic forms gain a foothold in it, much greater skill and knowledge are required to cope with the problem. Its solution is of the greatest economic importance to the gardener and to the greenhouse man.

Parasitic fungi, upon finding their way into a soil, do not necessarily interfere with the work of the beneficial bacteria, such as the ammonifiers and nitrifiers, for instance. Nor do they always influence the chemical or physical nature of the soil. Many of them directly attack the crop itself, causing serious diseases in the plants.

DAMPING OFF

This disease is very familiar to every grower of plants. It is peculiar to seedlings or tender plants, and is very prevalent in the greenhouse, the hot bed, the cold frame as well as in the field. It is induced by the presence of definite parasitic fungi, which thrive best in overwatered soils, and when the greenhouse is kept at a comparatively high temperature with poor ventilation. Damping off is also favoured by thick sowing and too much shade in the seed bed.

Symptoms of Damping Off. Every experienced grower knows the disease when he sees it. Seedlings freshly damped off are soft and water soaked at the base of the stem. If they are pulled they often break off easily. A more careful examination shows that the root system is entirely decayed, although the upper part of the stem and leaves may still be green, and also possibly fresh.

The degree of prostration in the seedlings is determined by the amount of moisture in the soil. If it is slight, the seedlings will become flabby and wilted before they topple over. With a high moisture content, they are more firm, but become prostrate as soon as infection sets in. The trouble usually begins in spots in the bed, thence spreading in every direction. Damping off is usually caused by several fungi, the chief of which is Pythium de Baryanum Hesse. The organism was first named and described by Hesse in 1874. Ward found it to be a very prevalent parasite in the garden soils of Europe. In America the fungus was first recognized as of great economic importance by Atkinson. The seedlings of most greenhouse plants may become subject to damping off by Pythium.

When examined under a compound microscope, Pythium de Baryanum is seen to be made up of coarse non-separate, highly granular, irregularly branched hyaline vegetative threads or mycelium. The younger growing threads are more finely granular. The oldest are coarsely granular or more often empty. These threads penetrate the cells of the host, where they obtain its food.

Pythium de Baryanum does not often fruit on the dead seedlings. The fruiting is better observed when the fungus is grown in pure culture. Under normal conditions it produces two forms of spores, conidia and oogonia. The summer spores, or conidia, are swellings formed at the tip of the hyphæ. These swellings readily break off from the mother threads and germinate by sending out a slender tube. This tube penetrates the seedling tissue where it grows and develops and after due incubation reproduces the disease.

The oospore, or sexual spore, is the stage which is most commonly found. The female organ (oogonium) first develops as a terminal enlargement which is cut off by a septum from the mother thread. Next or adjacent to it a slender tube is cut off from the mycelium by a septum. This tube performs the function of the male sexual organ and is known as antheridium. The latter then comes into close contact and empties all its content into the female oogonium. Fertilization thus takes place, and a mature egg, or oospore, or winter resting spore is formed.

The latest investigations have not yet disclosed whether or not Pythium de Baryanum is carried over from year to year by its oospores. It is apparently able to propagate itself indefinitely by its vegetative mycelium. Of the other fungi which are capable of producing a damping off in the greenhouse or seed bed may be mentioned Sclerotinia libertiana Fckl., Phoma solani Halst., Colletotrichum sp., Fusarium sp., Sclerotium rolfsii Sacc. and Rhizactonia solani Kuhn. Each of these, except the last, will be taken up separately in connection with the study of their respective hosts.

The fungi which cause damping off are introduced into the greenhouse, primarily with sick soil used in the compost, and also with infected manure. The practice of dumping diseased plants and all other infected material in the manure pile cannot be too strongly condemned. Sometimes very lightly infected plants with no visible symptoms of disease in the seed bed may nevertheless act as carriers, and like-wise infect the greenhouse soil.

DAMPING OFF OF CUTTINGS

Greenhouse men are often troubled with a damping off of cuttings. In specific cases this is brought about by parasitic fungi which, however, will be taken up at length under the discussion of the various hosts.

Root Rot

Although not so virulent as Pythium, Rhizoctonia is a frequent cause of considerable failure in green-house culture. The fungus causes a damping off of seedlings and cuttings and a serious root rot.

Symptoms: The symptoms of Rhizoctonia rot or wilt do not differ materially from those produced by Pythium de Baryanum. On older plants, however, Rhizoctonia produces cankers or deep lesions which are very characteristic. These are formed on the roots as well as on the base of the stem. The lesions are reddish brown and extend into the cortical or vital layer as well as into the woody tissue.

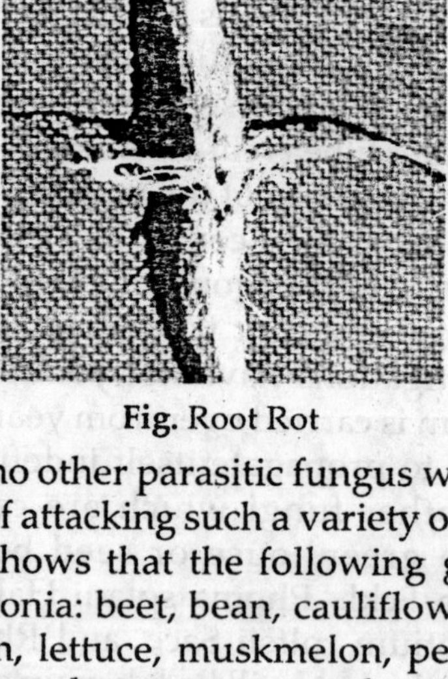

Fig. Root Rot

There is perhaps no other parasitic fungus which is so widespread and which is capable of attacking such a variety of hosts as Rhizoctonia. The work of Peltier shows that the following greenhouse crops are susceptible to Rhizoctonia: beet, bean, cauliflower, celery, cucumber, egg plant, horseradish, lettuce, muskmelon, pepper, radish, tomato, sweet alyssum, amaranthus, ornamental asparagus, china aster, begonia, candytuft, carnation, coleus, dianthus, lavatera, lobelia, pansy, poinsettia, sweet pea, violet. Cuttings of the following hosts are also reported by Peltier to damp off from Rhizoctonia: Abutilon hybridum, var. lavitzii, Acalypha wilkesiana, var. bicolour, A. wilkesiana, var.

tricolour; A. wilkesiana, var. marginata, Ageratum mexicanum, Alyssum odoratum; Coleus, Cuphea phatycentra, Tresine, Petunia, Piguerua trinervia, Lautolina chamoecyparissus, Sedum spectabile, Althernanthera, Vincia major.

The Organism. In the United States the first extended account of Rhizoctonia was given by Pammel. Many other excellent accounts by American workers have appeared from time to time, to which we shall have occasion to refer later. The genus Rhizoctonia includes several forms of sterile fungi, all of which are distinguished by the manner of growth in pure culture, and by its mycelium.

Young hyphae of R. solani Kuhn are at first hyaline, then deepening in colour from a yellowish to a deep brown. The young branches are somewhat narrowed at their point of union with the parent hyphæ and grow in a direction almost parallel with each other. A septum is also laid down at a short distance from the point of union with the parent mycelium.

There is another form of hypha which is made up of barrel shaped cells each of which is capable of germinating like a spore. In pure cultures R. solani produces sclerotia which are first soft, whitish, and which later become hard and dark. The fungus is, carried over from year to year as sclerotia which are able to withstand the effects of heat, cold, drought, or moisture.

Next in importance to Rhizoctonia is a group of fungi which belong to the genus Fusarium. Soils infected with these species of fungi become unfit for tomatoes, sweet peas, etc., thereby causing great financial losses to the greenhouse man. Individual difficulties will be taken up in studying each of these crops separately. As an illustration of a typical Fusarium sick soil let us consider the wilt of sweet pea. The cause of this trouble is a soil inhabiting fungus, Fusarium lath yri Taub.

Symptoms. The first symptom of the disease is a sudden flagging of the leaves, accompanied by general wilting and collapse of the seedling. Usually upon sowing the seeds a fair percentage germinate and reach the height of about 8 to 10 inches before they are attacked by the fungus. If the collapsed seedlings are allowed to remain on the ground, the stems will soon be covered with the sickle shaped spores. Eventually the decayed tissue rots and is soon invaded by small fruit flies which now begin to distribute the fungus from place to place by carrying its spores.

The Organism. The mycelium of Fusarium lath yri is hyaline, septate and branched. At an early age the mycelial cells round up into countless numbers of chlamydospores. Old cultures are practically one mass of these resting bodies. The spores are of two sorts, the macroconidia which are sickle shaped, 3-4 septate, the microconidia are one celled, minute spherical to elliptical.

Root Knot

Caused by Heterodera radicicola (Greef) Mull. Although root knot is most prevalent in light soils, it may sometimes be found in heavier lands. The trouble is most widespread out of doors in the South-ern States, where the winter is mild. In the North the worm is usually unable to winter over in the open unless it is protected by trash or dead weeds. It is, however, prevalent in greenhouses and is undoubtedly introduced with sick soil brought in from the field.

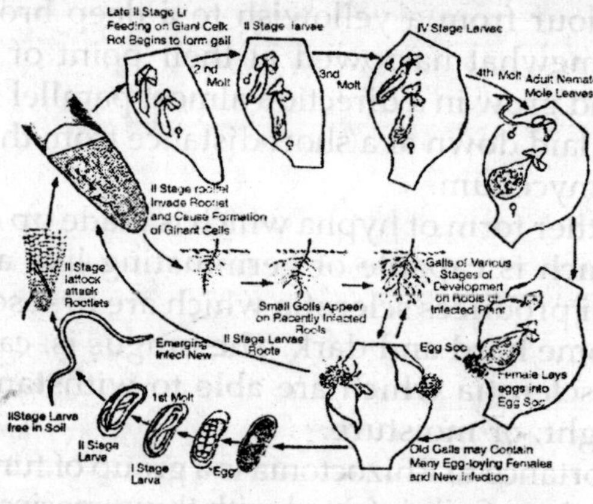

Fig. Root Knot

Symptoms. The disease is characterized by swellings or knots on the roots. These swellings may be variously shaped, and are often mistaken for the true nodules of legumes. Infected plants become stunted, pale, and usually linger for a long time before dying.

The Organism. The nematode is a very minute worm, seldom exceeding one twenty-fifth of an inch in length. It is semi-transparent so that it cannot be easily detected by the naked eye. In searching for the eel worm, it is necessary to break a fresh knot. Close examination will reveal two types of worms; a spindle shaped worm, the male, and a pearly white pear shaped organism, the female, firmly embedded in the gall tissue. The female is very prolific, depositing no less than 400 to goo eggs during her lifetime. The eggs are whitish, semi-transparent, bean shaped bodies, and too small to be noticed without the aid of a magnifying glass. The time which elapses until the eggs hatch depends largely upon weather conditions. In warm days the eggs hatch sooner than in cold days. Upon hatching, the young larvæ either remain in the tissue of the host plant in which they emerge, or, as is more often the case, leave the host and enter the soil. This is the only period during which the worms move about to any great extent in the soil, where they either remain for some length of time or immediately penetrate an-other root of the host. The nematodes in most cases become completely buried in the root tissue, establishing themselves in the soft cellular structure which is rich in food. The head of the worm is provided with a boring apparatus consisting of a sharply pointed spear, located in the mouth. This structure not only aids it in getting food but is also valuable in helping the young worms to batter through the cell walls before becoming definitely located.

The two sexes during the development are indistinguishable up to fifteen or twenty days, both being spindle shaped. In the molting or shedding of the skin, there is a marked change in the case of the female, especially in the posterior region of the body, which no longer possesses a tail-like appendage. Fertilization occurs soon after this molt, and many radical changes occur in the shape and structure of the organization of the worm.

The tertilized female increases rapidly in breadth and be-comes a pearly white flask or pear-shaped individual. At this stage it is far from being worm-like and may, therefore, be overlooked by one unfamiliar with the life-history of the eel worm. The young male is much like that of the young female larvæ, being spindle shaped in outline.

The male does not cause as much damage to the root tissue as the female, and its purpose in life seems to be only that of fertilizing the female, for after this function has been performed, it is quite probable that the male worm takes no more food. Omnivorous Nature of the Eel Worm. There are two hundred and thirty-five species of plants known to suffer from the eel worm. This number includes all the important families of the flowering plants. According to Bessey * the following are among the greenhouse plants subject to root knot: bean, beet, cantaloupe, cauliflower, cucumber, egg-plant, lettuce, radish, tomato.

LEAF BLIGHT NEMATODE

Beside the root knot disease which is caused by Heterodera radicicola there is another nematode which confines its injury to foliage only. Of the greenhouse hosts affected by this pest may be mentioned the Begonia, Asplenium nidus-avis, Pteris serrulata avistata, Pteris wimeseth, Pteris tremula, and Pelargonium.

Symptoms: On the Cincinnati begonia the symptoms, according to Clinton, are manifested as numerous small indistinct discolourations limited by the small veinlets. In time, however, the tiny spots enlarge and unite, forming a conspicuous reddish-brown blotch. Frequently infection is manifested as long streaks along the rnain veins. Often isolated spots occur in the midst of the surrounding healthy tissue.

On Asplenium nidus-avis, the trouble becomes conspicuous in dark brown areas from the base of the leaf near the midrib. These spread up-ward until the entire lower half of the leaf is killed. On Pteris, the spots appear as reddish brown bands reaching out from the midrib to the border, but limited sidewise by the small parallel cross veins.

The Organism. The nematode in question is a slender microscopical worm. The latter chooses the air chamber of the leaf in which to lay its eggs and upon hatching travels around in different parts of the same leaf or to the neighbouring foliage. The worm can travel only when there is a wet film on the leaves.

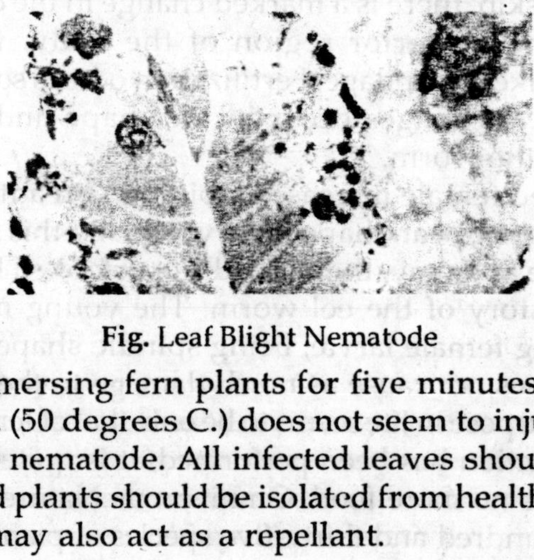

Fig. Leaf Blight Nematode

Control. Immersing fern plants for five minutes in water heated to 122 degrees F. (50 degrees C.) does not seem to injure the ferns, but seems to kill the nematode. All infected leaves should be cut off and burned. Diseased plants should be isolated from healthy ones. Spraying with Bordeaux may also act as a repellant.

SOIL

Soil, the loose material that covers the land surfaces of Earth and supports the growth of plants. In general, soil is an unconsolidated, or loose, combination of inorganic and organic materials. The inorganic components of soil are principally the products of rocks and minerals that have been gradually broken down by weather, chemical action, and other natural processes. The organic materials are composed of debris from plants and from the decomposition of the many tiny life forms that inhabit the soil.

Soils vary widely from place to place. Many factors determine the chemical composition and physical structure of the soil at any given location. The different kinds of rocks, minerals, and other geologic materials from which the soil originally formed play a role. The kinds of plants or other vegetation that grow in the soil are also important.

Topography—that is, whether the terrain is steep, flat, or some combination—is another factor. In some cases, human activity such as farming or building has caused disruption. Soils also differ in colour, texture, chemical makeup, and the kinds of plants they can support.

Soil actually constitutes a living system, combining with air, water, and sunlight to sustain plant life. The essential process of photosynthesis, in which plants convert sunlight into energy, depends on exchanges that take place within the soil. Plants, in turn, serve as a vital part of the food chain for living things, including humans. Without soil there would be no vegetation—no crops for food, no forests, flowers, or grasslands. To a great extent, life on Earth depends on soil.

The study of different soil types and their properties is called soil science or pedology. Soil science plays a key role in agriculture, helping farmers to select and support the crops on their land and to maintain fertile, healthy ground for planting. Understanding soil is also important in engineering and construction. Soil engineers carry out detailed analysis of the soil prior to building roads, houses, industrial and retail complexes, and other structures.

Soil takes a great deal of time to develop—thousands or even millions of years. As such, it is effectively a nonrenewable resource. Yet even now, in many areas of the world, soil is under siege. Deforestation, over-development, and pollution from humanmade chemicals are just a few of the consequences of human activity and carelessness. As the human population grows, its demand for food from crops increases, making soil conservation crucial.

Quality of Soil

The good or bad qualities of a soil have reference to the needs of the crops which are to be grown upon it, and it is only after a consideration of the requirements of plants that a clear conception can be formed of what characters the soil must possess for it to be a suitable medium on which healthy crops can be raised.

In the first place, soil, to be of any use, must be sufficiently loose and porous to allow the roots of plants to grow and extend freely. It may be so compact that root development is checked or stopped altogether, in which case the plant suffers. On the other hand it should not be too open in texture or the roots do not get a proper hold of the ground and are easily disturbed by wind: moreover such soils are liable to blow away, leaving the underground parts exposed to the air and drought.

The roots like all other parts of plants contain protoplasm or living material, which cannot carry on its functions unless it is supplied with an adequate amount of oxygen: hence the necessity for the continuous circulation of fresh air through the soil. If the latter is too compact or has its interstices filled with carbon dioxide gas or with water - as is the case when the ground is water-logged - the roots rapidly die of suffocation just as would an animal under the same conditions.

There is another point which requires attention. Plants need very considerable amounts of water for their nutrition and growth; the waterholding capacity is, therefore, important. If the soil holds too much it becomes water-logged and its temperature falls below the point for healthy growth, at any rate of the kinds of plants. Usually cultivated on farms and in gardens. If it allows of too free drainage drought sets in and the plants, not getting enough water for their needs, become stunted in size. Too much water is bad, and too little is equally injurious.

In addition, the temperature of the soil largely controls the yield of crops which can be obtained from the land. Soil whose temperature remains low, whether from its northerly aspect or from its high water content or other cause, is unsatisfactory, because the germination of seeds and the general life processes of plants cannot go on satisfactorily except at certain temperatures well above freezing-point.

A good soil should be deep to allow of extensive root development and, in the case of arable soils, easy to work with implements. Even when all the conditions above mentioned in regard to texture, water-holding capacity, aeration and temperature are suitably fulfilled the soil may still be barren: plant foodmaterial is needed. This is usually present in abundance although it may not be available to the plant under certain circumstances, or may need to be replenished or increased by additions to the soil of manures or fertilizers.

CHIEF CONSTITUENTS OF THE SOIL

An examination of the soil shows it to be composed of a vast number of small particles of sand, clay, chalk and humus, in which are generally imbedded larger or smaller stones. It will be useful to consider the nature of the four chief constituents just mentioned and their bearing upon the texture, water-holding capacity and other characters which were referred to in the previous section.

Sand consists of grains of quartz or flint, the individual particles of which are large enough to be seen with the unaided eye or readily felt as gritty grains when rubbed between the finger and thumb. When a little soil is shaken up with water in a tumbler the sand particles rapidly fall to the bottom and form a layer which resembles ordinary sand of the seashore or river banks. Chemically pure sand is silicon dioxide (SiO_2) or quartz, a clear transparent glass-like mineral, but as ordinarily met with, it is more or less impure and generally coloured reddish or yellowish by oxide of iron.

A soil consisting of sand entirely would be very loose, would have little capacity to retain water, would be liable to become very hot in the daytime and cool at night and would be quite unsuitable for growth of plants. The term clay is often used by chemists to denote hydrated silicate of alumina ($Al_2O_3.2SiO_2.2H_2O$), of which kaolin or china clay is a fairly pure form. This substance is present in practically all soils but in comparatively small amounts.

Even in the soils which farmers speak of as stiff clays it is rarely present to the extent of more than I or 2%. The word " clay " used in the agricultural sense denotes a sticky intractable material which is found to consist of exceedingly fine particles (generally less than. 005 mm. in diameter) of sand and other minerals derived from the decomposition of rocks, with a small amount of silicate of alumina. The peculiar character which clay possesses is probably due not to its chemical composition but to its physical state.

When wet it becomes sticky and almost impossible to move or work with farm implements; neither air nor water can penetrate freely. In a dry state it becomes hard and bakes to a brick. It holds water well and is consequently cold, needing the application of much heat to raise its temperature. It is obvious, therefore, that soil composed entirely of clay is as useless as pure sand so far as the growth of crops upon it is concerned. Chalk consists, when quite pure, of calcium carbonate (CaC03), a white solid substance useful in small amounts as a plant foodmaterial, though in excess detrimental to growth. Alone, even when broken up into small pieces, it is unsuitable for the growth of plants.

Humus, the remaining constituent of soil, is the term used for the decaying vegetable and animal matter in the soil. A good illustration of it is peat. Its water-holding capacity is great, but it is often acid, and when dr y it is light and incapable of supporting the roots of plants properly. Few of the commonly cultivated crops can live in a soil consisting mainly of humus.

From the above account it will be understood that not one of the four chief soil constituents is in itself of value for the growth of crops, yet when they are mixed, as they usually are in the soils met with in nature, one corrects the deficiencies of the other. A perfect soil would be such a blend of sand, clay, chalk and humus as would contain sufficient clay and humus to prevent drought, enough sand to render it pervious to fresh air and prevent waterlogging, chalk enough to correct the tendency to acidity of the humus present, and would have within it various substances which would serve as food-materials to the crops.

COMPOSITION OF SOILS

Soils comprise a mixture of inorganic and organic components: minerals, air, water, and plant and animal material. Mineral and organic particles generally compose roughly 50 percent of a soil's volume. The other 50 percent consists of pores—open areas of various shapes and sizes. Networks of pores hold water within the soil and also provide a means of water transport. Oxygen and other gases move through pore spaces in soil. Pores also serve as passageways for small animals and provide room for the growth of plant roots.

Inorganic Material

The mineral component of soil is made up of an arrangement of particles that are less than 2.0 mm (0.08in) in diameter. Soil scientists divide soil particles, also known as soil separates, into three main size groups: sand, silt, and clay.

According to the classification scheme used by the United States Department of Agriculture (USDA), the size designations are: sand, 0.05 to 2.00 mm (0.002 to 0.08 in); silt 0.002 to 0.05 mm (0.00008 to 0.002 in); and clay, less than 0.002 mm (0.00008 in). Depending upon the rock materials from which they were derived, these assorted mineral particles ultimately release the chemicals on which plants depend for survival, such as potassium, calcium, magnesium, phosphorus, sulfur, iron, and manganese.

Organic Material

Organic materials constitute another essential component of soils. Some of this material comes from the residue of plants, the remains of plant roots deep within the soil, or materials that fall on the ground, such as leaves on a forest floor. These materials become part of a cycle of decomposition and decay, a cycle that provides important nutrients to the soil. In general, soil fertility depends on a high content of organic materials.

Even a small area of soil holds a universe of living things, ranging in size from the fairly large to the microscopic: earthworms, mites, millipedes, centipedes, grubs, termites, lice, springtails, and more. And even a gram of soil might contain as many as a billion microbes—bacteria and fungi too small to be seen with the naked eye. All these living things form a complex chain: Larger creatures eat organic debris and excrete waste into the soil, predators consume living prey, and microbes feed on the bodies of dead animals. Bacteria and fungi, in particular, digest the complex organic compounds that make up living matter and reduce them to simpler compounds that plants can use for food. A typical example of bacterial action is the formation of ammonia from animal and vegetable proteins.

Other bacteria oxidize the ammonia to form nitrogen compounds called nitrites, and still other bacteria act on the nitrites to form nitrates, another type of nitrogen compound that can be used by plants. Some types of bacteria are able to fix, or extract, nitrogen directly from the air and make it available in the soil.Ultimately, the decay of plant and animal material results in the formation of a dark-coloured organic matter known as humus. Humus, unlike plant residues, is generally resistant to further decomposition.

CHEMICAL COMPOSITION OF THE SOIL

It has been found by experiment that plants need for their nutritive process and their growth, certain chemical elements, namely, carbon, hydrogen, oxygen, nitrogen, sulphur, phosphorus, potassium, magnesium, calcium and iron. With the exception of the carbon and a small proportion of the oxygen and nitrogen, which may be partially derived from the air, these elements are taken from the soil by crops. The following table shows the amounts of the chief constituents removed by certain crops in lb per acre: - Plants also remove from the soil silicon, sodium, chlorine, and other elements which are, nevertheless, found to be unessential for the growth and may therefore be neglected here.

Leguminous crops take some of the nitrogen which they require from the air, but most plants obtain it from the nitrates present in the soil. The sulphur exists in the soil chiefly in the form of sulphates of magnesium, calcium and other metals; the phosphorus mainly as phosphates of calcium, magnesium and iron; the potash, soda and other bases as silicates and nitrates; calcium and magnesium carbonates are also common constituents of many soils. In the ordinary chemical analyses of the soil determinations are made of the nitrogen and various carbonates present as well as of the amount of phosphoric acid, potash, soda, magnesia and other components soluble in strong hydrochloric acid.

Soil Formation

Soil formation is an ongoing process that proceeds through the combined effects of five soil-forming factors: parent material, climate, living organisms, topography, and time. Each combination of the five factors produces a unique type of soil that can be identified by its characteristic layers, called horizons. Soil formation is also known as *pedogenesis* (from the Greek words *pedon*, for "ground," and *genesis*, meaning "birth" or "origin").

Parent Material

The first step in pedogenesis is the formation of parent material from which the soil itself forms. Roughly 99 percent of the world's soils derive from mineral-based parent materials that are the result of weathering, the physical disintegration and chemical decomposition of exposed bedrock.

The small percentage of remaining soils derives from organic parent materials, which are the product of environments where organic matter accumulates faster than it decomposes. This accumulation can occur in marshes, bogs, and wetlands.

Bedrock itself does not directly give rise to soil. Rather, the gradual weathering of bedrock, through physical and chemical processes, produces a layer of rock debris called regolith. Further weathering of this debris, leading to increasingly smaller and finer particles, ultimately results in the creation of soil.

In some instances, the weathering of bedrock creates parent materials that remain in one place. In other cases, rock materials are transported far from their source—blown by wind, carried by moving water, and borne inside glaciers.

Climate

Climate directly affects soil formation. Water, ice, wind, heat, and cold cause physical weathering by loosening and breaking up rocks. Water in rock crevices expands when it freezes, causing the rocks to crack.

Rocks are worn down by water and wind and ground to bits by the slow movement of glaciers. Climate also determines the speed at which parent materials undergo chemical weathering, a process in which existing minerals are broken down into new mineral components. Chemical weathering is fastest in hot, moist climates and slowest in cold, dry climates.

Climate also influences the developing soil by determining the types of plant growth that occur. Low rainfall or recurring drought often discourage the growth of trees but allow the growth of grass. Soils that develop in cool rainy areas suited to pines and other needle-leaf trees are low in humus.

Living Organisms

As the parent material accumulates, living things gradually gain a foothold in it. The arrival of living organisms marks the beginning of the formation of true soil. Mosses, lichens, and lower plant forms appear first.

As they die, their remains add to the developing soil until a thin layer of humus is built up. Animals' waste materials add nutrients that are used by plants. Higher forms of plants are eventually able to establish themselves as more and more humus accumulates. The presence of humus in the upper layers of a soil is important because humus contains large amounts of the elements needed by plants.

Living organisms also contribute to the development of soils in other ways. Plants build soils by catching dust from volcanoes and deserts, and plants' growing roots break up rocks and stir the developing soil. Animals also mix soils by tunneling in them.

WATER IN THE SOIL

The importance of an adequate supply of water to growing crops cannot well be over-estimated. During the life of a plant there is a continuous stream of water passing through it which enters by the root-hairs in the soil and after passing along the stem is given off from the 'stomata of the leaves into the open air above ground. It has been estimated that an acre of cabbage will absorb from the land and transpire from its leaves more than ten tons of water per day when the weather is fine.

In addition to its usefulness in maintaining a turgid state of the young cells without which growth cannot proceed, water is itself a plant food-material and as absorbed from the soil contains dissolved in it all the mineral food constituents needed by plants for healthy nutrition. Without a sufficient supply plants remain stunted and the crop yield is seriously reduced, as we see in dry seasons when the rainfall is much below the average.

If one condition is more necessary than another for good crops it is a suitable supply of water, for no amount of manuring or other treatment of the soil will make up for a deficient rainfall. The amount needed for the most satisfactory nutrition varies with different plants. In the case of fair average farm crops it has been shown that for the production of one ton of dry matter contained in them from 300 to 500 tons of water has been absorbed and utilized by the plants.

This may be more than the rainfall, in which case irrigation or special control of the water supply may be necessary. The water-holding capacity of a soil depends upon the amount of free space between the particles of which it is composed into which water can enter. In most cases this amounts to from 30 to 50% of the volume of the soil. When the pore-space of the soil is filled with water it becomes water-logged and few plants can effect absorption by their roots under such conditions.

The root-hairs die from want of air, and the whole plant soon suffers. Fields of wheat and other cereals rarely recover after a week's submergence, but orchards and many trees when at rest in winter withstand a flooded or water-logged condition of the soil for two or three weeks without damage.

The most satisfactory growth is maintained when the amount of water present is not more than 40 to 60% of what would saturate it. Under such conditions each particle of soil is surrounded by a thin film of water and in the pore-space air can freely circulate. It is from such films that the root-hairs absorb all that plants require for their growth.

The movement of water into the root-hairs is brought about by the osmotic action of certain salts in their cell-sap. Crops are, however, unable to absorb all the water present in the soil, for when the films become very thin they are held more firmly or cling with more force to the soil particles and resist the osmotic action of the root-hairs. Plants have been found to wither and die in sandy soils containing i a % of water, and in clay soils in which there was still present 8% of water.

When a long glass tube open at both ends is filled with soil and one end is dipped in a shallow basin of water, the water is found to move upwards through the soil column just as oil will rise in an ordinary lamp wick. By this capillary action water may be transferred to the upper layers of the soil from a depth of several feet below the surface. In this manner plants whose roots descend but a little way in the ground are enabled to draw on deep supplies.

Not only does water move upwards, but it is transferred by capillarity in all directions through the soil. The amount and speed of movement of water by this means, and the distance to which it may be carried, depend largely upon the fineness of the particles composing the soil and the spaces left between each.

The ascent of water is most rapid through coarse sands, but the height to which it will rise is comparatively small. In clays whose particles are exceedingly minute the water travels very slowly but may ultimately reach a height of many feet above the level of the " water-table " below. While this capillary movement of water is of great importance in supplying the needs of plants it has its disadvantages, since water may be transferred to the surface of the soil, where it evapo rates into the air and is lost to the land or the crop growing upon it.. The loss in this manner was found to be in one instance over a pound. of water per day per square foot of surface, the " water-table " being about 4 or 5 ft. below. One of the most effective means of conserving soil moisture is by " mulching," *i.e.* by covering the surface of the soil with some loosely compacted material such as straw, leaf-refuse or stablemanure.

The space between the parts of such substances is too large to admit of capillary action; hence the water conveyed to the surface of the soil is prevented from passing upwards any further except by slow evaporation through the mulching layer. A loose layer of earth spread over the surface of the soil acts in the same way, and a similarly effective mulch may be prepared by hoeing the soil, or stirring it to a depth of one or two inches with harrows or other implements. The hoe and harrow are therefore excellent tools for use in dry weather. Rolling the land is beneficial to young crops. in dry weather, since it promotes capillary action by reducing the soil spaces. It should, however, be followed by a light hoeing or harrowing.

In the semi-arid regions of the United States, Argentina and other countries where the average annual rainfall lies between ioa to 20 in., irrigation is necessary to obtain full crops every year. Good crops, however, can often be grown in such areas without irrigation if attention is paid to the proper circulation of water in the soil and means for retaining it or preventing excessive loss by evaporation. Of course care must be exercised in the selection of plants - such as sorghum, maize, wheat, and alfalfa or lucerne - which are adapted to dry conditions and a warm climate.

So far as the water-supply is concerned - and this is what ultimately determines the yield of crops - the rain which falls upon the soil should be made to enter it and percolate rapidly through its interstices. A deep porous bed in the upper layers is essential, and this should consist of fine particles which lie close to each other without any tendency to stick together and " puddle " after heavy showers. Every effort should be made to prepare a good mealy tilth by suitable ploughing, harrowing and consolidation.

In the operation of ploughing the furrow slice is separated from the soil below, and although in humid soils this layer may be left to settle by degrees, in semi-arid regions this loosened layer becomes. dry if left alone even for a few hours and valuable water evaporates. into the air. To prevent this various implements, such as disk harrows and specially constructed rollers, may be used to consolidate the upper stirred portion of the soil and place it in close capillary relationship with the lower unmoved layer. If the soil is allowed to become dry and pulverized, rain is likely to run off or " puddle the surface without penetrating it more than a very short distance. Constant hoeing or harrowing to maintain a natural soil mulch layer of 2 or 3 in. deep greatly conserves the soil water below. In certain districts where the rainfall is low a crop can only be obtained once every alternate year, the intervening season being devoted to tillage with a view of getting the rain into the soil and retaining it there for the crop in the following year.

Soil scientists also characterize soils according to how effectively they retain and transport water. Once water enters the soil from rain or irrigation, gravity comes into play, causing water to trickle downward. Water is also taken up in great quantities by the roots of plants: Plants use anywhere from 200 to 1,000 kg (440 to 2,200 lb) of water in the formation of 1 kg (2.2 lb) of dry matter. Soils differ in their capacity to retain moisture against the pull exerted by gravity and by plant roots. Coarse soils, such as those consisting of mostly of sand, tend to hold less water than do soils with finer textures, such as those with a greater proportion of clays.

Water also moves through soil pores independently of gravity. This movement can occur via capillary action, in which water molecules move because they are more attracted to the pore walls than to one another. Such movement tends to occur from wetter to drier areas of the soil. The movement from soil to plant roots can also depend on how tightly water molecules are bound to soil particles.

The attraction of water molecules to each other is an example of cohesion. The attraction of water molecules to other materials, such as soil or plant roots, is a type of adhesion. These effects, which determine the so-called matric potential of the soil, depend largely on the size and arrangement of the soil particles. Another factor that can affect water movement is referred to as the osmotic potential.

The osmotic potential hinges on the amount of dissolved salts in the soil. Soils high in soluble salt tend to reduce uptake of water by plant roots and seeds. The sum of the matric and osmotic potentials is called the total water potential. In soil, water carries out the essential function of bringing mineral nutrients to plants. But the balance between water and air in the soil can be delicate.

An overabundance of water will saturate the soil and fill pore spaces needed for the transport of oxygen. The resulting oxygen deficiency can kill plants. Fertile soils permit an exchange between plants and the atmosphere, as oxygen diffuses into the soil and is used by roots for respiration. In turn, the resulting carbon dioxide diffuses through pore spaces and returns to the atmosphere. This exchange is most efficient in soils with a high degree of porosity. For farmers, gardeners, landscapers, and others with a professional interest in soil health, the process of aeration—making holes in the soil surface to permit the exchange of air—is a crucial activity. The burrowing of earthworms and other soil inhabitants provides a natural and beneficial form of aeration.

Bacteria in the Soil

Recent science has made much progress in the investigation of the micro-organisms of the soil. Whereas the soil used to be looked upon solely as a dead, inert material containing certain chemical substances which serve as food constituents. of the crops grown upon it, it is now known to be a place of habitation for myriads of minute living organisms upon whose activity much of its fertility depends.

They are responsible for many important chemical processes which make the soil constituents more available and better adapted to the nutrition of crops. One cubic centimetre of soil taken within a foot or so from the surface contains from II to 2 millions of bacteria of many different kinds, as well as large numbers of fungi. In the lower depths of the soil the numbers. decrease, few being met with at a depth of 5 or 6 ft.

The efficiency of many substances, such as farm-yard manure,. guanos, bone-meal and all other organic materials, which are spread over or dug or ploughed into the land for the benefit of farm and' garden crops, is bound up with the action of these minute living beings. Without their aid most manures would be useless for plant growth. Farm-yard manure, guanos and other fertilizers. undergo decomposition in the soil and become broken down into compounds of simple chemical composition better suited for absorption by the roots of crops, the changes involved being directly due to the activity of bacteria and fungi.

Much of the work carried on by these organisms is not clearly understood; there are, however, certain processes which have been extensively investigated and to these it is necessary to refer. It has been found by experiment that the nitrogen needed by practically all farm crops except leguminous ones is best supplied in the form of a nitrate; the rapid effect of nitrate of soda when used' as a top dressing to wheat or other plants is well known to farmers. It has long been known that when organic materials such as the dung and urine of animals, or even the bodies of animals and plants, are applied to the soil, the nitrogen within them becomes oxidized, and ultimately appears in the form of nitrate of lime, potash or some other base. The nitrogen in decaying roots, in the dead stems. and leaves of plants, and in humus generally is sooner or later changed into a nitrate, the change being effected by bacteria. That: the action of living organisms is the cause of the production of nitrates is supported by the fact that the change does not occur when the soil is heated nor when it is treated with disinfectants which destroy or check the growth and life of bacteria.

The process resulting in the formation of nitrates in the soil is spoken of as nitrification. The steps in the breaking down of the highly complex nitrogenous proteid compounds contained in the humus of the soil, or applied to the latter by the farmer in the form of dung and organic refuse generally, are many and varied; most frequently the insoluble proteids are changed by various kinds of putrefactive bacteria into soluble proteids (peptones, &c.), these into simpler amido-bodies, and these again sooner or later into compounds 7of ammonia. The urea in urine is also rapidly converted by the uro-bacteria into ammonium carbonate. The compounds of ammonia thus formed from the complex substances by many varied kinds of micro-organisms are ultimately oxidized into nitrates.

The change takes place in two stages and is effected by two special groups of nitrifying bacteria, which are present in all soils. In the first stage the ammonium compounds are oxidized to nitrites by the agency of very minute motile bacteria belonging to the genus Nitrosomonas. The further oxidation of the nitrite to a nitrate is effected by bacteria belonging to the genus Nitrobacter. Several conditions must be fulfilled before nitrification can occur. In the first place an adequate temperature is essential; at 5° or 6° C. (40-43° F.) the process is stopped, so that it does not go on in winter. In summer, when the temperature is about 24° C. (75° F.), nitrification proceeds at a rapid rate.

The organisms do not carry on their work in soils deficient in air; hence the process is checked in water-logged soils. The presence of a base such as lime or magnesia (or their carbonates) is also essential, as well as an adequate degree of moisture: in dry soils nitrification ceases. It is the business of the farmer and gardener to promote the activity of these organisms by good tillage, careful drainage and occasional application of lime to soils which are deficient in this substance.

It is only when these conditions are attended to that decay and nitrification of dung, guano, fish-meal, sulphate of ammonia and other manures take place, and the constituents which they contain become available to the crops for whose benefit they have been applied to the land. Nitrates are very soluble in water and are therefore liable to be washed out of the soil by heavy rain. They are, however, very readily absorbed by growing plants, so that in summer, when nitrification is most active, the nitrates produced are usually made use of by crops before loss by drainage takes place. In winter, however, and in fallows loss takes place in the subsoil water.

There is also another possible source of loss of nitrates through the activity of denitrifying bacteria. These organisms reduce nitrates to nitrites and finally to ammonia and gaseous free nitrogen which escapes into the atmosphere. Many bacteria are known which are capable of denitrification, some of them being abundant in fresh dung and upon old straw. They can, however, only carry on their work extensively under anaerobic conditions, as in waterlogged soils or in those which are badly tilled, so that there is but little loss of nitrates through their agency.

An important group of soil organisms are now known which have the power of using the free nitrogen of the atmosphere for the formation of the complex nitrogenous compounds of which their bodies are largely composed. By their continued action the soil becomes enriched with nitrogenous material which eventually through the nitrification process becomes available to ordinary green crops. This power of " fixing nitrogen," as it is termed, is apparently not possessed by higher green plants.

The bacterium, *Clostridium pasteurianum,* common in most soils, is able to utilize free nitrogen under anaerobic conditions, and an organism known as *Azotobacter chroococcum* and some others closely allied to it, have similar powers which they can exercise under aerobic conditions. For the carrying on of their functions they all need to be supplied with carbohydrates or other carbon compounds which they obtain ordinarily from humus and plant residues in the soil, or possibly in some instances from carbohydrates manufactured by minute green algae with which they live in close union.

Certain bacteria of the nitrogenfixing class enter into association with the roots of green plants, the best-known examples being those which are met with in the nodules upon the roots of clover, peas, beans, sainfoin and other plants belonging to the leguminous order. That the fertility of land used for the growth of wheat is improved by growing upon it a crop of beans or clover has been long recognized by farmers. The knowledge of the cause, however, is due to modern investigations. When wheat, barley, turnips and similar plants are grown, the soil upon which they are cultivated becomes depleted of its nitrogen; yet after a crop of clover or other leguminous plants the soil is found to be richer in nitrogen than it was before the crop was grown. This is due to the nitrogenous root residues left in the land. Upon the roots of leguminous plants characteristic swollen nodules or tubercles are present. These are found to contain large numbers of a bacterium termed *Bacillus radicicola* or *Pseudomonas radicicola*. The bacteria, which are present in almost all soils, enter the root-hairs of their host plants and ultimately stimulate the production of an excrescent nodule, in which they live. For a time after entry they multiply, obtaining the nitrogen necessary for their nutrition and growth from the free nitrogen of the air, the carbohydrate required being supplied by the pea or clover plant in whose tissues they make a home.

The nodules increase in size, and analysis shows that they are exceedingly rich in nitrogen up to the time of flowering of the host plant. During this period the bacteria multiply and most of them assume a peculiar thickened or branched form, in which state they are spoken of as bacteroids.

Later the nitrogen-content of the nodule decreases, most of the organisms, which are largely composed of proteid material, becoming digested and transformed into soluble nitrogenous compounds which are conducted to the developing roots and seeds. After the decay of the roots some of the unchanged bacteria are left in the soil, where they remain ready to infect a new leguminous crop.

The nitrogen-fixing nodule bacteria can be cultivated on artificial media, and many attempts have been made to utilize them for practical purposes. Pure cultures may be made and after dilution in water or other liquid can be mixed with soil to be ultimately spread over the land which is to be infected.

The method of using them most frequently adopted consists in applying them to the seeds of leguminous plants before sowing, the seed being dipped for a time in a liquid containing the bacteria. In this manner organisms obtained from red clover can be grown and applied to the seed of red clover; and similar inoculation can be arranged for other species, so that an application of the bacteria most suited to the particular crop to be cultivated can be assured.

In many cases it has been found that inoculation, whether of the soil or of the seed, has not made any appreciable difference to the growth of the crop, a result no doubt due to the fact that the soil had already contained within it an abundant supply of suitable organisms.

But in other instances greatly increased yields have been obtained where inoculation has been practised. More or less pure cultures of the nitrogen-fixing bacteria belonging to the *Azotobacter* group have been tried and recommended for application to poor land in order to provide a cheap supply of nitrogen. The application of pure cultures of bacteria for improving the fertility of the land is still in an experimental stage. There is little doubt, however, that in the near future means will be devised to obtain the most efficient work from these minute organisms, either by special artificial cultivation and subsequent application to the soil, or by improved methods of encouraging their healthy growth and activity in the land where they already exist.

Improvement of Soils

The fertility of a soil is dependent upon a number of factors, some of which, such as the addition of fertilizers or manures, increase the stock of available food materials in the soil while others, such as application of clay or humus, chiefly influence the fertility of the land by improving its physical texture.

The chief processes for the improvement of soils are:liming, claying and marling, warping, paring and burning, and green manuring. Most of these more or less directly improve the land by adding to it certain plant food constituents which are lacking, but the effect of each process is in reality very complex. In the majority of cases the good results obtained are more particularly due to the setting free of " dormant " or " latent " food constituents and to the amelioration of the texture of the soil, so that its aeration, drainage, temperature and water-holding capacity are altered for the better.

The material which chemists call calcium carbonate is met with in a comparatively pure state in chalk. It is present in variable amounts in limestones of all kinds, although its white ness may there be masked by the presence of iron oxide and other coloured substances. Carbonate of lime is also a constituent to a greater or lesser extent in almost all soils. In certain sandy soils and in a few stiff clays it may amount to less than 4%, while in others in limestone and chalk districts there may be 50 to 80% present.

Pure carbonate of lime when heated loses 44% of its weight, the decrease being due to the loss of carbon dioxide gas. The resulting white product is termed calcium oxide lime, burnt lime, quicklime, cob lime, or caustic lime. This substance absorbs and combines with water very greedily, at the same time becoming very hot, and falling into a fine dry powder,' calcium hydroxide or slaked lime, which when left in the open slowly combines with the carbon dioxide of the air and becomes calcium carbonate, from which we began.

When recommendations are made about liming land it is necessary to indicate more precisely than is usually done which of the three classes of material named above - chalk, quicklime or slaked lime - is intended. Generally speaking the oxide or quicklime has a more rapid and greater effect in modifying the soil than slaked lime, and this again greater than the carbonate or chalk.

Lime in whatever form it is applied has a many-sided influence in the fertility of the land. It tends to improve the tilth and the capillarity of the soil by binding sands together somewhat and by opening up clays. If applied in too great an amount to light soils and peat land it may do much damage by rendering them too loose and open.

The addition of small quantities of lime, especially in a caustic form, to stiff greasy clays makes them much more porous and pliable. A lump of clay, which if dried would become hard and intractable, crumbles into pieces when dried after adding to it 2% of lime. The lime causes the minute separate particles of clay to flocculate or group themselves together into larger compound grains between which air and water can percolate more freely.

It is this power of creating a more crumbly tilth on stiff clays that makes lime so valuable to the farmer. Lime also assists in the decomposition of the organic matter or humus in the soil and promotes nitrification; hence it is of great value after green manuring or where the land contains much humus from the addition of bulky manures such as farm-yard dung.

This tendency to destroy organic matter makes the repeated application of lime a pernicious practice, especially on land which contains little humus to begin with. The more or less dormant nitrogen and other constituents of the humus are made immediately available to the succeeding crop, but the capital of the soil is rapidly reduced, and unless the loss is replaced by the addition of more manures the land may become sterile. Although good crops may follow the application of lime, the latter is not a direct fertilizer or manure and is no substitute for such. Its best use is obtained on land in good condition, but not where the soil is poor.

When used on light dry land it tends to make the land drier, since it destroys the humus which so largely assists in keeping water in the soil. Lime is a base and neutralizes the acid materials present in badly drained meadows and boggy pastures. Weeds, therefore, which need sour conditions for development are checked by liming and the better grasses and clovers are encouraged. It also sets free potash and possibly other useful plant food-constituents of the soil. Liming tends to produce earlier crops and destroys the fungus which causes finger-and-toe or club-root among turnips and cabbages.

Land which contains less than about z /c, of lime usually needs the addition of this material. The particular form in which lime should be applied for the best results depends upon the nature of the soil. In practice the proximity to chalk pits or lime kilns, the cost of the lime and cartage, will determine which is most economical. Generally speaking light poor lands deficient in organic matter will need the less caustic form or chalk, while quicklime will be most satisfactory on the stiff clays and richer soils.

On the stiff soils overl y ing the chalk it was formerly the custom to dig pits through the soil to the rock below. Shafts 20 or 30 ft. deep were then sunk, and the chalk taken from horizontal tunnels was brought to the surface and spread on the land at the rate of about 60 loads per acre. Chalk should be applied in autumn, so that it may be split by the action of frost during the winter.

Quicklime is best applied, perhaps, in spring at the rate of one ton per acre every six or eight years, or in larger doses-4 to 8 tons - every 15 to 20 years. Small dressings applied at short intervals give the most satisfactory results. The quicklime should be placed in small heaps and covered with soil if possible until it is slacked and the lumps have fallen into powder, after which it may be spread and harrowed in. Experiments have shown that excellent effects can be obtained by applying 5 or 6 cwt. of ground quicklime.

Gas-lime is a product obtained from gasworks where quicklime is used to purify the gas from sulphur compounds and other objectionable materials. It contains a certain amount of unaltered caustic lime and slacked lime, along with sulphates and sulphides of lime, some of which have an evil odour. As some of these sulphur compounds have a poisonous effect on plants, gas-lime cannot be applied to land directly without great risk or rendering it incapable of growing crops of any sort - even weeds - for some time.

It should therefore be kept a year or more in heaps in some waste corner and turned over once or twice so that the air can gain access to it and oxidize the poisonous ingredients in it. Many soils of a light sandy or gravelly or peaty nature and liable to drought and looseness of texture can be improved by the addition of large amounts of clay of an ordinary character.

Similarly soils can be improved by applying to them marl, a substance consisting of a mixture of clay with variable proportions of lime. Some of the chalk marls, which are usually of a yellowish or dirty grey colour, contain clay and 50 to 80% of carbonate of lime with a certain proportion of phosphate of lime. Such a material would not only have an influence on the texture of the land but the lime would reduce the sourness of the land and the phosphate of lime supply one of the most valuable of plant foodconstituents.

The beneficial effects of marls may also be partially due to the presence in them of available potash. Typical clay-marls are tenacious, soapy clays of yellowish-red or brownish colour and generally contain less than 50% of lime. When dry they crumble into small pieces which can be readily mixed with the soil by ploughing. Many other kinds of marls are described; some are of a sandy nature, others stony or full of the remains of small shells.

The amount and nature of the clay or marl to be added to the soil will depend largely upon the original composition of the latter, the lighter sands and gravel requiring more clay than those of firmer texture. Even stiff soils deficient in lime are greatly improved in fertility by the addition of marls.

In some cases as little as 40 loads per acre have been used with benefit, in others 180 loads have not been too much. The material is dug from neighbouring pits or sometimes from the fields which are to be improved, and applied in autumn and winter. When dry and in a crumbly state it is harrowed and spread and finally ploughed in and mixed with the soil.

On some of the strongest land it was formerly the practice to add to and plough into it burnt clay, with the object of making the land work more easily. The burnt clay moreover carried *Cl ay* with it potash and other materials in a state readily available to the crops. The clay is dug from the land or from ditches or pits and placed in heaps of 60 to ioo loads each, with faggot wood, refuse coals or other fuel.

Great care is necessary to prevent the heaps from becoming too hot, in which case the clay becomes baked into hard lumps of brick-like material which cannot be broken up.

With careful management, however, the clay dries and bakes, becoming slowly converted into lumps which readily crumble into a fine powder, in which state it is spread over and worked into the land at the rate of 40 loads per acre.

The paring and burning of land, although formerly practised as an ordinary means of improving the texture and fertility of arable fields, can now only be looked upon as a practice p to be adopted for the purpose of bringing rapidly into cultivation very foul leys or, land covered with a coarse turf. The practice is confined to poorer types of land, such as heaths covered with furze and bracken or fens and clay areas smothered with rank grasses and sedges.

To reduce such land to a fit state for the growth of arable crops is very difficult and slow without resort to paring and burning. The operation consists of paring off the tough sward to a depth of I to 2 in. just sufficient to effectually damage the roots of the plants forming the sward and then, after drying the sods and burning them, spreading the charred material and ashes over the land. The turf is taken off either with the breast plough - a paring tool pushed forward from the breast or thighs by the workman - or with specially constructed paring ploughs or shims. The depth of the sod removed should not be too thick or burning is difficult and too much humus is destroyed unnecessarily, nor should it be too thin or the roots of the herbage are not effectually destroyed.

The operation is best carried out in spring and summer. After being pared off the turf is allowed to dry for a fortnight or so and is then placed in small heaps a yard or two wide at the base, a little straw or wood being put in the middle of each heap, which is then lighted. As burning proceeds, more turf is added to the outside of the heaps in such a manner as to allow little access of air. Every care should be taken to burn and char the sod thoroughly without permitting the heap to blaze. The ashes should be spread as soon as possible and covered by a shallow ploughing. The land is then usually sown with some rapidly growing green crop, such as rape, or with turnips.

Paring and burning improves the texture of clay lands, particularly if draining is carried out at the same time. It tends to destroy insects and weeds, and gets rid of acidity of the soil. No operation brings old turf into cultivation so rapidly. Moreover the beneficial effects are seen in the first crop and last for many years.

Many of the mineral plant food-constituents locked up in the coarse herbage and in the upper layers of the soil are made immediately available to crops. The chief disadvantage is the loss of nitrogen which it entails, this element being given off into the air in a free gaseous state. It is best adapted for application to clays and fen lands and should not be practised on shallow light sands or gravelly soils, since the humus so necessary for the fertility of such areas is reduced too much and the soil rendered too porous and liable to suffer from drought.

Many thousands of acres of low-lying peaty and sandy land adjoining the tidal rivers which flow into the Humber have been improved by a process termed " warping." The warp consists of fine muddy sediment which is suspended in the tidal river water and appears to be derived from material scoured from the bed of the Humber by the action of the tide and a certain amount of sediment brought down by the tributary streams which join the Humber some distance from its mouth. The field or area to be warped must lie below the level of the water in the river at high tide. It is first surrounded by an embankment, after which the water from the river is allowed to flow through a properly constructed sluice in its bank, along a drain or ditch to the land which is prepared for warping. By a system of carefully laid channels the water flows gently over the land, and deposits its warp with an even level surface.

At the ebb of the tide the more or less clear water flows back again from the land into the main river with sufficient force to clean out any deposit which may have accumulated in the drain leading to the warped area, thus allowing free access of more warpladen water at the next tide. In this manner poor peats and sands may be covered with a large layer of rock soil capable of growing excellent crops.

The amount of deposit laid over the land reaches a thickness of two or three feet in one season of warping, which is usually practised between March and October, advantage being taken of the spring tides during these months.

The new warp is allowed to lie fallow during the winter after being laid out in four-yard " lands " and becomes dry enough to be sown with oats and grass and clover seeds in the following spring. The clover-grass ley is then grazed for a year or two with sheep, after which wheat and potatoes are the chief crops grown on the land.

Green manures are crops which are grown especially for the purpose of ploughing into the land in a green or actively growing state. The crop during its growth obtains a considerable amount of *Green.* carbon from the carbon dioxide of the air, and builds it up *Manuring.* i nto compounds which when ploughed into the land become humus.

The carbon compounds of the latter are of no direct nutritive value to the succeeding crop, but the decaying vegetable tissues very greatly assist in retaining moisture in light sandy soils, and in clay soils also have a beneficial effect in rendering them more open and allowing of better drainage of superfluous water and good circulation of fresh air within them. The ploughing-in of green crops is in many respects like the addition of farm-yard manure.

Their growth makes no new addition of mineral food-constituents to the land, but they bring useful substances from the subsoil nearer to the surface, and after the decay of the buried vegetation these become available to succeeding crops of wheat or other plants. Moreover, where deep-rooting plants are grown the subsoil is aerated and rendered more open and suitable for the development of future crops.

The plants most frequently used are white mustard, rape, buckwheat, spurry, rye, and several kinds of leguminous plants, especially vetches, lupins and serradella. By far the most satisfactory crops as green manures are those of the leguminous class, since they add to the land considerable amounts of the valuable fertilizing constituent, nitrogen, which is obtained from the atmosphere. By nitrification this substance rapidly becomes available to succeeding crops. On the light, poor sands of Saxony Herr Schultz, of Lupitz, made use of serradella, yellow lupins and vetches as green manures for enriching the land in humus and nitrogen, and found the addition of potash salts and phosphates very profitable for the subsequent growth of potatoes and wheat. He estimated that by using leguminous crops in this manner for the purpose of obtaining cheap nitrogen he reduced the cost of production of wheat more than 50%.

The growing crops should be ploughed in before flowering occurs; they should not be buried deeply, since decay and nitrification take place most rapidly and satisfactorily when there is free access of air to the decaying material. When the crop is luxuriant it is necessary to put a roller over it first, to facilitate proper burial by the plough.

The best time for the operation appears to be late summer and autumn. (J. PE.) *Soil and Disease. - The* influence of different kinds of soil as a factor in the production of disease requires to be considered, in regard not only to the nature and number of the microorganisms they contain, but also to the amount of moisture and air in them and their capacity for heat.

The moisture in soil is derived from two sources—the rain and the ground-water. Above the level of the ground-water the soil is kept moist by capillary attraction and by evaporation of the water below, by rainfall, and by movements of the ground-water; on the other hand, the upper layers are constantly losing moisture by evaporation from the surface and through vegetation. When the ground-water rises it forces air out of the soil; when it falls again it leaves the soil moist and full of air.

The nature of the soil will largely influence the amount of moisture which it will take up or retain. In regard to water, all soils have two actions - namely, permeability and absorbability. Permeability is practically identical with the speed at which percolation takes place; through clay it is slow, but increases in rapidity through marls, loams, limestones, chalks, coarse gravels and fine sands, reaching a maximum in soil saturated with moisture.

The amount of moisture retained depends mainly upon the absorbability of the soil, and as it depends largely on capillary action it varies with the coarseness or fineness of the pores of the soil, being greater for soils which consist of fine particles. The results of many analyses show that the capacity of soils for moisture increases with the amount of organic substances present; decomposition appears to be most active when the moisture is about 4%, but can continue when it is as low as 2%, while it appears to be retarded by any excess over 4%. Above the level of the ground-water all soils contain air, varying in amount with the degree of looseness of the soil. Some sands contain as much as 50% of air of nearly the same composition as atmospheric air. The oxygen, however, decreases with the depth, while the carbon dioxide increases.

SOIL CHARACTERISTICS

Scientists can learn a lot about a soil's composition and origin by examining various features of the soil. Colour, texture, aggregation, porosity, ion content, and pH are all important soil characteristics.

Colour

Soils come in a wide range of colours—shades of brown, red, orange, yellow, gray, and even blue or green. Colour alone does not affect a soil, but it is often a reliable indicator of other soil properties. In the surface soil horizons, a dark colour usually indicates the presence of organic matter. Soils with significant organic material content appear dark brown or black. The most common soil hues are in the red-to-yellow range, getting their colour from iron oxide minerals coating soil particles. Red iron oxides dominate highly weathered soils. Soils frequently saturated by water appear gray, blue, or green because the minerals that give them the red and yellow colours have been leached away.

Texture

A soil's texture depends on its content of the three main mineral components of the soil: sand, silt, and clay. Texture is the relative percentage of each particle size in a soil. Texture differences can affect many other physical and chemical properties and are therefore important in measures such as soil productivity.

Soils with predominantly large particles tend to drain quickly and have lower fertility. Very fine-textured soils may be poorly drained, tend to become waterlogged, and are therefore not well-suited for agriculture. Soils with a medium texture and a relatively even proportion of all particle sizes are most versatile. A combination of 10 to 20 percent clay, along with sand and silt in roughly equal amounts, and a good quantity of organic materials, is considered an ideal mixture for productive soil.

Aggregation

Individual soil particles tend to be bound together into larger units referred to as aggregates or soil peds. Aggregation occurs as a result of complex chemical forces acting on small soil components or when organisms and organic matter in soil act as glue binding particles together. Soil aggregates form soil structure, defined by the shape, size, and strength of the aggregates.

There are three main soil shapes: platelike, in which the aggregates are flat and mostly horizontal; prismlike, meaning greater in vertical than in horizontal dimension; and blocklike, roughly equal in horizontal and vertical dimensions and either angular or rounded. Soil peds range in size from very fine—less than 1 mm (0.04 in)—to very coarse—greater than 10 mm (0.4 in). The measure of strength or grade refers to the stability of the structural unit and is ranked as weak, moderate, or strong. Very young or sandy soils may have no discernible structure.

Porosity

The part of the soil that is not solid is made up of pores of various sizes and shapes—sometimes small and separate, sometimes consisting of continuous tubes. Soil scientists refer to the size, number, and arrangement of these pores as the soil's porosity. Porosity greatly affects water movement and gas exchange.

Well-aggregated soils have numerous pores, which are important for organisms that live in the soil and require water and oxygen to survive. The transport of nutrients and contaminants will also be affected by soil structure and porosity.

Ion Content

Soils also have key chemical characteristics. The surfaces of certain soil particles, particularly the clays, hold groupings of atoms known as ions. These ions carry a negative charge. Like magnets, these negative ions (called anions) attract positive ions (called cations).

Cations, including those from calcium, magnesium, and potassium, then become attached to the soil particles, in a process known as cation exchange. The chemical reactions in cation exchange make it possible for calcium and the other elements to be changed into water-soluble forms that plants can use for food. Therefore, a soil's cation exchange capacity is an important measure of its fertility.

pH

Another important chemical measure is soil pH, which refers to the soil's acidity or alkalinity. This property hinges on the concentration of hydrogen ions in solution. A greater concentration of hydrogen results in a lower pH, meaning greater acidity. Scientists consider pure water, with a pH of 7, neutral. The pH of a soil will often determine whether certain plants can be grown successfully. Blueberry plants, require acidic soils with a pH of roughly 4 to 4.5. Alfalfa and many grasses, on the other hand, require a neutral or slightly alkaline soil. In agriculture, farmers add limestone to acid soils to neutralize them.

SOIL USE

For most of human history, soil has not been treated as the valuable and essentially nonrenewable resource that it is. Erosion has devastated soils worldwide as a result of overuse and misuse. In recent years, however, farmers and agricultural experts have become increasingly concerned with soil management.

Erosion

Erosion is the wearing away of material on the surface of the land by wind, water, or gravity. In nature, erosion occurs very slowly, as natural weathering and geologic processes remove rock, parent material, or soil from the land surface. Human activity, on the other hand, greatly increases the rate of erosion. In the United States, the farming of crops accounts for the loss of over 3 billion metric tons of soil each year.

In a cultivated field from which crops have been harvested, the soil is often left bare, without protection from the elements, particularly water. Raindrops smash into the soil, dislodging soil particles. Water then carries these particles away. This movement may take the form of broad overland flows known as sheet erosion.

More often, the eroding soil is concentrated into small channels, or rills, producing so-called rill erosion. Gravity intensifies water erosion. Landslides, in which large masses of water-loosened soil slide down an incline, are a particularly extreme example.

Wind erosion occurs where soils are dry, bare, and exposed to winds. Very small soil particles can be suspended in the air and carried away with the wind. Larger particles bounce along the ground in a process called saltation.

Soil Management

To prevent exposure of bare soil, farmers can use techniques such as leaving crop residue in the soil after harvesting or planting temporary growths, such as grasses, to protect the soil from rain between crop-growing seasons. Farmers can also control water runoff by planting crops along the slope of a hill (on the contour) instead of in rows that go up and down.

Soil faces many threats throughout the world. Deforestation, overgrazing by livestock, and agricultural practices that fail to conserve soil are three main causes of accelerated soil loss. Other acts of human carelessness also damage soil. These include pollution from agricultural pesticides, chemical spills, liquid and solid wastes, and acidification from the fall of acid rain.

Loss of green spaces, such as grassland and forested areas, in favour of impermeable surfaces, such as pavement, buildings, and developed land, reduces the amount of soil and increases pressure on what soil remains. Soil is also compacted by heavy machinery and off-road vehicles. Compaction rearranges soil particles, increasing the density of the soil and reducing porosity. Crusts form on compacted soils, preventing water movement into the soil and increasing runoff and erosion.

With the world's population now numbering upwards of 6 billion people—a figure that may rise to 10 billion or more within three decades—humans will depend more than ever on soil for the growth of food crops.

Yet the rapidly increasing population, the intensity of agriculture, and the replacement of soil with concrete and buildings all reduce the capacity of the soil to fulfill this need. As a result of an increased awareness of soil's importance, many changes are being made to protect soil. Recent interest in soil conservation holds the promise that humanity will take better care of this precious resource.

POOR SEEDLING EMERGENCE

Patchy crops reduce yield and profit. Poor seedling emergence is more of a problem under minimum tillage, because there is less soil disturbance and more stubble retention. It is less of a problem under conventional tillage, because the seedbeds are finely worked and the seeds are planted into bare ground. However, rather than reject minimum tillage because of its problems, it is preferable to address those problems.

POSSIBLE CAUSES

Poor seedling emergence may be due to:

- Poor seed–soil contact
- Inaccurate seed placement
- A soil temperature that is too low or too high
- Soil insects or soil-borne disease
- Surface crusting after sowing
- Poor quality seed.

Poor seedling vigour and establishment after emergence may caused by the presence of compacted soil beneath the seedling. The soil may have already compacted before sowing, or the planting tine may have smeared the bottom and sides of the seed trench.

Planting Machinery

If poor seedling emergence is due to poor seed–soil contact or inaccurate seed placement, then adjusting your sowing machinery correctly will help to improve emergence.

Surface Crusting

Rain after sowing and before seedling emergence may form a surface crust that reduces emergence. Once seedlings have emerged, the problem of surface crusting is less. However, until the crop reaches full ground cover (and protects the surface from raindrop impact) surface crusting will continue to reduce infiltration.

All soils seal under raindrop impact—even self-mulching soils seal if they are bare. On some soils, the seal is weak or it cracks as it dries; it offers little resistance to seedling emergence. However, on other soils, the seal forms a crust on drying, and this crust prevents seedling emergence.

Further light rain will help emergence by softening the surface; further heavy rain will hinder emergence by strengthening the surface crust. Clay soils can crust if the surface is sodic. The treatment for such soils is gypsum.

COMPACTION MANAGEMENT

A small degree of compaction can be beneficial. Soil that is too loose may not be good for plant growth: a loose seedbed dries out too quickly and gives poor contact between the seeds and the soil; seeds may not germinate. The optimum level of compaction depends on the soil types, the plant, and the irrigation management during the growing season. Too much compaction can cause yield reduction. To reduce soil compaction:

- Time mechanical operations carefully
- Reduce axle loads
- Keep livestock off wet cropping country
- Confine traffic to laneways
- Use low ground pressure (wide) tyres (with caution).

The method of compaction repair will depend on the soil's capabilities. A cracking clay that swells and shrinks behaves quite differently to a hardsetting, fine, sandy or silty soil. If a soil swells and shrinks on wetting and drying you can use that natural action to break up a compacted layer. If a hardpan has developed in a soil with little clay, you will need to consider other options.

Example:

- Biological repair—that is, using growing plants to break up a compacted layer or increasing the soil organic matter content
- Mechanical repair—that is, deep ripping or mouldboard ploughing.

BIOLOGICAL REPAIR

Biological repair means using growing plants to restore soil structure. It usually involves a yield penalty. The plants will suffer in restoring the soil structure, and will produce a lower yield. Repair of compaction should therefore work in with the season, the market prices, and your long-term plans.

Biological repair includes:

- biological ripping (using plants to dry and crack the soil)
- Biological drilling (using tap-rooted plants to 'drill' through acompacted layer)
- Increasing soil organic matter content.

Biological Ripping

The easiest way to restore soil structure in compacted cracking clays is to dry the soil, using any plants with a vigourous root system. Drying causes the soil to shrink and crack, allowing water, air and roots to penetrate through the cracks. The process is called biological ripping.

Biological Drilling of Non-Cracking Soils

In soils that don't crack, such as sandy, silty or loamy soils, 'biological ripping' is not an option. However, 'biological drilling' is possible. A strong tap-rooted plant such as lucerne can force its way through a compacted layer and leave root channels for subsequent plant roots to use. This is 'biological drilling'. The compacted layer must be moist: no root will enter dry soil. Depending upon how badly compacted the subsoil is, some roots will not be able to penetrate the compacted layer and the crop will suffer.

Increasing Soil Organic Matter

Some soils, such as fine sandy and silty soils, depend heavily on organic matter for the maintenance of their structures. Even cracking soils benefit from increased organic matter content. Increasing the organic matter of a soil content is a slow process. A pasture phase incorporated into your vegetable crop rotation is perhaps the most costeffective way to achieve this.

MECHANICAL REPAIR

Deep ripping and mouldboard ploughing are sometimes useful for quick results, but they are expensive operations. The initial success of the operation depends on the moisture content of the soil and the depth of tillage. Continuing success depends on managing traffic to avoid having machinery on the soil when it is too wet.

The Importance of Moisture Content

Soil moisture is important when you are working soil. A soil (particularly a clay) that is worked too wet will compact, while a silty or fine sandy soil that is worked too dry may powder. In both instances the soil will then structurally degrade. Before deep ripping, dig a hole to see if there is a compacted layer. If there is, check the layer's depth, thickness, and moisture content.

The severity of compaction will also influence your decision on deep ripping. If there is no compacted layer, there is no need to deep rip. Deep ripping is expensive, and it could do more harm than good if the soil is too wet—a costly way of creating soil structural damage where none existed before! A compacted layer must be at the right moisture content to benefit from the ripping operation.

When you are cultivating clays (or when the compacted layer is clay) the layer must be dry enough to shatter, rather than smear. When the compacted layer is other than clay, some moisture is desirable so that the layer does not powder.

The depth and thickness of the compacted layer indicates how deep to set the ripper tine. To be effective, the tine should work just below the compacted layer. Dig a hole after a short run and check the effectiveness of the ripping operation.

The upper surface of the fruiting body is dull gray-brown and smooth. The lower surface has elongate openings (pores) or split-like openings (gills). These fruiting bodies produce millions of tiny spores which may, in some cases, serve to spread the decay fungus to other areas. Also, surface molds, "mildews," and stain fungi are often found growing on the surface of damp wood and can be confused with decay fungi. Although these organisms may discolour the wood, they do not break down wood fibres and thus do not weaken its structure.

However, these organisms indicate that moisture is present and that decay will likely proceed if a wood-rotting fungus becomes established in the wood.

Chapter 4

Nematodes

Wilt of cotton resembles the wilt disease of tomato, cowpeas, watermelons, cabbage, and several other crops. It is primarily a disease of the water-conducting vessels of the woody or stem part of the plant and is caused by a fungus that inhabits the soil. The disease is complicated by nematodes, the eellike microscopic worms that also inhabit the soil and provide the openings through which the wilt fungus enters cotton roots.

Nematodes reduce the root growth and increase the susceptibility of cotton plants to fusarium wilt. Thus the disease is considered a wilt-nematode complex. Some nematodes enter the tip end of young roots. Others enter and feed on the root tissue some distance back from the tip. The root knot nematode makes galls, or knots, on roots, which later decay and leave the ends of the water-conducting vessels open and exposed to soil-borne organisms. The meadow nematode, feeding some distance from the root tips, may cause a pruning off of the small rootlets.

The Wounds made by those and other species of nematode provide numerous openings for the wilt fungus, Fusarium oxysporum f. vasinfectum. Once the wilt pathogen gets into the vascular system, which conducts water to all parts of the plant, it can grow and spread throughout the woody portion. In the vascular ducts it may be found in pure culture, and appears especially adapted to grow in this tissue while most other organisms from the soil are excluded.

The wilt disease of cotton is distributed throughout the world wherever American cottons are grown in acid alluvial sandy soils. Pathologists believe the disease originated in Mexico or Central America in the same locality as upland cotton. The wilt fungus is carried inside cotton seeds and has been transported by the seed to new cotton-growing areas.

In the United States the disease occurs in all States from Virginia to eastern Oklahoma and Texas. It is limited in those States by low rainfall and alkaline soils. Greatest losses occur in Coastal Plain soils of the Carolinas, Georgia, and Alabama. Wilt is also an important disease in Arkansas, Louisiana, Mississippi, and southeastern Texas. Although the disease is more severe in the sandy soils, it occurs in the lighter soils scattered throughout some of the heavier soils series of the Piedmont and Mississippi Delta regions. Losses to the cotton crop from the wilt-nematode disease complex have been greater than those caused by any other disease except possibly Texas root rot.

Losses in yields result from reduced stands, stunted plants, small bolls, and poor-quality lint. Before the development, distribution, and general use of wilt-resistant varieties, losses to individual growers often amounted to 75 to90 per cent of the crop. The growing of sea-island cotton had ceased on many acres before 1902, when Rivers, a resistant variety, became available.

The first wilt-resistant upland variety, Dillon, was released in 1905. Estimates of crop losses beginning in 1920 indicate wilt losses from 1 to 5 per cent for the different States. Additional losses from the root knot nematode ranged from a trace to 3 per cent. Between 1940 and 1950 the release of improved wilt-resistant varieties brought further reductions in Wilt losses. Now the losses from nematodes and wilt together probably do not exceed 3 or 4 per cent in any State.

Symptoms of wilt may appear on cotton plants at any stage of development. The earliest symptoms to be seen on seedlings and small plants are the yellowing and browning of cotyledons and leaves. The affected parts ultimately die and fall off. The bare stem soon blackens and dies. The first symptom in older plants may be stunting, followed by yellowing, wilting, and dropping of most of the leaves.

Leaf discoloration first appears near the margin of the blade near a vein. The affected areas enlarge, and an abscission layer may form at the base of the petiole, causing the leaf to drop. An outstanding symptom is the browning and blackening of the woody tissue. When a stem or branch is cut crosswise, the discoloration is usually found in a ring just beneath the bark.

Sometimes the discoloration is dispersed through the woody cylinder. In advanced cases, discoloration may extend throughout the plant from the roots through the stem, branches, leaf petioles, and peduncles and into the bolls.

Wilting mostly occurs gradually, but after a rain, following a dry period, plants may wilt suddenly and in large numbers. Wilted plants may produce some bolls, which usually are smaller and open prematurely. Plants may die one at a time until the stand is reduced or largely eliminated, depending on the susceptibility of the variety and the amount of infestation. The wilt fungus survives in the soil on organic matter. It grows as a threadlike mycelium and produces two kinds of spores. One type, known as conidia, is relatively short-lived. The second type, chlamydospores, is a resting stage, which may live longer. Both types give rise to a mycelium that infects the roots of the host plants.

Dissemination by spores probably accounts for the rapid spread of the disease to all parts of a field once the disease is introduced by cotton seed. Spores may be washed about in the field or blown about by the wind or transported by many other means. The wilt fungus lives in the soil indefinitely once it is introduced even though cotton and other susceptible plants are not grown. Fields not planted to cotton for as long as 25 years have shown severe wilting the first year after cotton planting was resumed. No method of eradicating the organism economically from fields is known.

Many laboratory experiments indicate that with large amounts of inoculum the fungus enters healthy cotton roots in the absence of openings made by nematodes. Field experiments show that the openings caused by nematodes largely account for the infections occurring naturally.

In experiments at the Alabama Agricultural Experiment Station, I found that wilt is readily controlled with soil fumigants, which reduce or eliminate nematodes before planting. After entering the small roots, the fungus inhabits the water-conducting vessels and spreads by growth of the mycelium and by movement of spores upward in the water stream. The vascular tubes become browned and later blackened by the formation of gumlike substances and by growth of tyloses. Plugging or partial plugging of the vessels lowers the flow of water and uptake of salts from the soil and stunts the plant or causes wilting. Toxic products, which injure the host cells, are also produced by the fungus.

Browning, drying, and killing follow; the ultimate falling of most leaves is a symptom largely produced by toxic materials. When bolls are present, the mycelium may grow through the peduncle into the seed. After the plant dies, the organism invades all its tissues; if enough moisture is present, spores are produced, which may be spread to all parts of fields.

The root knot Nematode is the most common of the nematodes that affect cotton roots. It occurs on cotton roots in all lighter soils of the Cotton Belt. The immature larvae infest cotton plants by invasion through the soft root tips. After entering the tip, the larvae push their way between the cells. Then they become stationary and feed by puncturing all the cell walls within reach with a spearlike stylet and sucking out the juices from inside the cell. The affected cells grow much larger and proliferate to form knots, or galls. With susceptible varieties and the feeding of large numbers of nematodes, the knotlike enlargements may become a half-inch in diametre.

Tissues in the galls are quite soft and are likely to decay and leave the ends of vascular bundles exposed to the wilt fungus. The enlarged worms are filled with eggs, which hatch and release numerous young nematodes. They, in turn, feed on any new cotton rootlets in the vicinity. Some cotton plants become infected by wilt when quite young; perhaps the mycelium enters the root tissue with the young larvae as well as later when the galls decompose.

If nematodes have made many points of entry for the fungus, multiple infections gradually envelop the root system, even in fusarium-resistant plants of upland cotton. The meadow or root rot nematode (Pratylenchus pratensis) multiplies rapidly on corn, crabgrass, and other fibrous rooted crops, following which it might become the predominant species on cotton. The meadow nematode is also found abundantly in some soils too heavy in texture to support the root knot nematode.

It may enter the soft cortical root tissue at any point near the growing zone. In feeding, the female moves about, destroying cells and depositing eggs throughout a short segment of the root. The young larvae intensify the destruction of the cortical tissues so that usually the rootlet is cut off. Many openings are thus left for the wilt organism to enter directly into the vascular tubes.

Direct losses in yield from nematodes alone may be of greater importance in some soils than losses resulting from the wilt-nematode complex. Serious losses occur in some irrigated soils where fusarium wilt is unknown. Losses occur throughout the lighter soils of the rain belt but often go unnoticed by growers.

The main effect of nematodes is a reduced rate of growth, which means smaller plants and lower yields. Growers are inclined to attribute them to other causes. The effect of nematodes on yield can best be demonstrated by using soil fumigants: Often they double yields in severely infested fields in both the western irrigated and eastern rain belt soils.

Nematode-reducing rotations also demonstrate how great are the losses caused by nematodes. In some western soils the root knot nematode becomes the limiting factor to profitable cotton production and rotation or soil fumigation becomes necessary. The host range of the cotton wilt fungus is limited. It may enter the roots of a number of different crops, but it produces wilt symptoms only on cotton, okra, coffeeweed (Cassia tora), and some varieties of burley tobacco. The host range of the root knot nematode is extensive. It attacks and reproduces in the roots of more than 1,200 different species of plants.

Studies made by V. H. Young and W. H. Tharp at the Arkansas Agricultural Experiment Station show that the maximum growth rate of the fungus and maximum disease development require relatively high soil temperature and moisture. Soil temperatures between 80 and 90 F. were favourable for maximum disease development. Although some wilt occurs in seedlings and small plants in the field in April and May, maximum appearance of symptoms is in late June or July and August when soil temperatures are highest. In greenhouse studies, maximum wilt development was obtained at 80 to 90 per cent of the water-holding capacity of the soil.

High soil temperatures also are favourable for nematode development. The soil moisture conditions that are favourable for plant growth usually are favourable for nematodes. Field observations indicate that maximum wilt development occurs during years of highest soil temperatures, which also favour maximum nematode development.

High soil moisture tends to lower soil temperatures. Alternate dry, hot periods of rather long duration followed by rains provide conditions favourable for maximum wilt development. Measures to control wilt consist of practices directed at both the wilt organism and nematodes. Resistant varieties and the use of balanced fertilisers to produce healthy plants help control Fusarium. The planting of varieties resistant to root knot, the use of rootknot reducing crops in rotation with cotton, and the application of soil fumigants are important in the control of nematodes.

The early history of the wilt disease illustrates the early participation of the Department of Agriculture in the study and control of a disease threatening a major farm crop. Cotton wilt was first described by George F. Atkinson at the Alabama Agricultural Experiment Station in 1892. Atkinson also clarified the effects of wilt, which farmers confused with rust or potash deficiency.

Atkinson also described the root knot nematode on cotton and found that it provided openings through which the wilt fungus entered cotton roots. The Department of Agriculture was requested to help on the wilt problem in 1895, when the disease became serious on the sea-island plantations off the coast of South Carolina. Erwin F. Smith visited the area at that time and began his studies on the wilt diseases of cotton, watermelon, and cowpeas. E. L. Rivers, a sea-island plantation operator, in 1895 began the selection work that ultimately led to the production of the wilt-resistant sea-island variety that bears his name. W. A. Orton, of the Department of Agriculture, entered the work in South Carolina in 1899. He cooperated in the later breeding phases and distribution of the Rivers variety.

Orton also initiated a successful breeding programme with upland cottons in 1899. The early breeding work of Rivers and Orton was the forerunner of the breeding programmes that ultimately led to the practical control of cotton wilt and the wilts of other crops. The importance of wilt-resistant varieties was demonstrated by the release of Rivers. It saved the industry that centreed on the islands off the South Carolina coast and extended along the coast.

The early work by Rivers and Orton also established the plant-to-row selection method essentially as it is used by modern breeders. They picked out individual plants that had survived on severely infested soils. The following year the seed from each plant was planted in a single row. The rows that combined the best resistance and agronomic characters were then increased to establish the variety. Orton extended this method to include hybridisation to improve upland varieties in work that he and associates initiated in 1899. The method was adopted by other workers in breeding flax, peas, and watermelons resistant to fusarium wilt.

The inheritance of fusarium resistance in sea-island cotton later was found to be determined by two dominant pairs of factors. The resistance in sea-island, when pure, approaches immunity. Such high resistance greatly simplified the problem of selecting for resistance; it was fortunate that the earliest breeding work was done with sea-island. Upland cotton does not possess factors for the same high degree of resistance and has not yet been bred to compare with the resistance of Rivers sea-island.

From the breeding of wilt-resistant upland varieties initiated by Orton and his associates came Dillon, which was developed from a wilt-resistant selection of Jackson's Limbless in South Carolina and released in 1905. They developed a second variety, Dixie, in Alabama. A cross between Dixie and Triumph yielded a third variety, Dixie-Triumph, which was widely grown and used extensively in later selection and crossing.

Beginning in 1913 the Alabama Agricultural Experiment Station made available to growers Cook 307 and other related varieties, which were developed by H. B. Tisdale. Additional varieties released before 1940 were Cleveland, Miller, Express, Lightning Express, Super Seven, Clevewilt, Toole, and Lewis 63. Most of them matured rather late and yielded less than the best nonresistant kinds. The long effort of breeders to produce agronomically superior wilt-resistant varieties was climaxed in 1942 with the introduction of Coker too wilt. It was as productive as nonresistant varieties and had other desirable characteristics such as a medium staple length, earliness, and adaptability to a wide range of growing conditions. It quickly became popular in southeastern and south central regions where wilt is a factor. Other superior wilt-resistant varieties, Empire, Stone-wilt, Pandora, White Gold, and Plains, were introduced in 1940-1950.

A moderate amount of root knot resistance is associated with fusarium resistance in varieties developed in the Southeast and selected on wilt- and nematode-infested soils. In the process of selecting for fusarium resistance and high yield in segregating progenies; moderate resistance to root knot is obtained without specific effort on the part of the breeder.

This resistance contributes to the fusarium resistance and to the yield of such lines. High resistance to root knot is not present in any upland varieties. However, there are considerable differences between the most susceptible and the most resistant ones. Varieties combining both fusarium resistance and moderate root knot resistance are Stonewilt, Coker too wilt, and Plains. Several varieties considered susceptible to root knot are Rowden, Miller, Deltapine, Bobshaw, Stoneville, Empire, and Pandora.

A search for high resistance to root knot was made at the Alabama Agricultural Experiment Station in 1951. A number of the world cotton species and many types and varieties were planted, and the roots were later examined for sise and abundance of galls. Several sources of resistance were found, the most promising of which were Gossypium barbadense var. darwinii and two wild cottons from Mexico.

The latter two introductions from Mexico are much like upland (G. hirsutum) but are not productive. These three were crossed and back-crossed to productive upland varieties. The segregating progenies will be selected for root knot resistance and further backcrossed. Eventually high root knot resistance in combination with desirable agronomic characters will be obtained. Improved root knot resistance will increase the yields of varieties grown on infested soils and should also improve the resistance to fusarium wilt.

The use of balanced fertilisers to maintain a vigorous growth of cotton plants is an important cultural practice to prevent losses from cotton Wilt. Potash tends to reduce wilt losses. Nitrogen and phosphorus tend to increase wilt within certain limits. The proper balance of nitrogen, phosphorus, and potash gives the maximum yields and best control of wilt when no one of the three elements is deficient.

and best control of wilt when no one of the three elements is deficient. Cotton rust, or potash deficiency, was early confused with cotton wilt. George F. Atkinson discovered that ample application of potash prevented the occurrence of rust. Rust on cotton continued to cause serious losses and increased the losses from wilt until 1926, when American sources of potash were developed, which provided ample supplies at a reasonable cost.

Nematocides can control fusarium wilt and nematodes and thereby can double yields of resistant varieties and triple and quadruple the yields of susceptible varieties grown on soils heavily infested by both pathogens. In experiments in Alabama in 1947, I learned that wilt is controlled indirectly by destroying the nematodes. The yields are increased as a result of control of both the wilt and nematodes. On treated plots better stands are maintained, the plants grow off more rapidly, grow larger, and produce greater yields.

Most economical control is obtained by applying the material in the row at the rate of 6 to 8 gallons an acre a few days before planting. Materials giving satisfactory control are ethylene dibromide and a mixture of dichloropropene and dichloropropane. Their trade names are Dowfume W-40 and D D. Row application gives control of nematodes in a zone near the young plants. Once the plants are established and growing rapidly, later nematode infections do little damage to the crop. Annual applications are necessary if only the row application is made.

Rotating Cotton with crops that reduce the amount of root knot is a cultural practice of value in reducing wilt losses and increasing yields. Crops that reduce root knot nematodes are grasses, sorghum, small grains, corn, peanuts, crotalaria, velvetbeans, alfalfa, and nematode-resistant cowpeas. Several crops that are particularly susceptible to root knot tend to intensify wilt and nematode losses. Among them are cowpeas susceptible to root knot, annual lespedezas, sweet-potatoes, and tobacco.

Crops grown during mild winters for green manure sometimes increase populations of root knot to the extent of causing increased losses in the cotton crop that follows them. Blue lupine, Austrian winter peas, and vetch are examples. On the contrary, increased organic matter is beneficial in improving yields and reducing wilt losses.

Rotations help to control other diseases of cotton. An example is ascochyta blight of young plants, which sometimes becomes epidemic. Rotation is likewise of value against bacterial blight infections in the Southwest and irrigated valleys, where plant refuse is not decayed over winter. The rotation of cotton with other crops has not become a widely used practice. Farmers are reluctant to follow some other crops with cotton, particularly peanuts and hay crops, which exhaust the soils of some minerals to the extent that yields of cotton may be reduced. Other reasons may have to do with the accessibility of the land to the house, weeds, and the tendency to select the most productive land for cotton.

The Rot That Attacks 2,000 Species. Cotton root rot, caused by the soil-inhabiting fungus Phymatotrichum omnivorum, flourishes on more than 2,000 species of wild and cultivated plants. Cotton grows in most of the area where the fungus is present, and the disease is usually designated as cotton root rot, although it also attacks the roots of trees, shrubs, fruits, and vegetables. The monocots grasses, corn, sorghum, and such are considered immune to it.

The fungus occurs naturally in the alkaline soils of the southwestern United States and northern Mexico. It has not been found in other parts of the world. The disease is serious in parts of Texas, Oklahoma, New Mexico, Arizona, and northern Mexico. It has been reported in California, Nevada, Utah, Arkansas, and Louisiana. It is particularly damaging in the blackland prairie of central Texas.

It has been alarmingly destructive. In Arizona, the average yearly loss between 1924 and 1929 was 10.3 per cent. Losses in Texas were estimated at 130,000 bales in 1918, 314,000 in 1919, 630,000 in 1920, 444,000 in 1928, 300,000 in 1937, and 191,000 in 1939. Those were reductions in yields of 5 to 15 per cent.

The losses fluctuate from year to year according to acreage in cotton and the moisture conditions under which the crop is grown. High temperature and high soil moisture favour the development of the disease. Symptoms of infection are sometimes evident 6 to 8 weeks after planting, but ordinarily are most apparent during the period of squaring and fruiting.

At first only an occasional plant z"may display symptoms, but within days or weeks the surrounding plants succumb. By midseason and later, irregular or circular spots or patches of brown, dead plants present a characteristic symptom of the presence of the disease. Such spots shade off abruptly from the dead, browned plants through a narrow zone of wilted plants into normal green plants.

The first above-ground symptom on a plant is a slight yellowing or bronzing of the leaves. There is an increase in leaf temperature, which one can feel. Wilting of the leaves occurs in a day or two and is rapidly followed by drying and browning of the foliage. The browned leaves stay on the plant for some time. Roots of infected cotton plants show various symptoms, depending on the length of time the root has been subjected to the action of the parasite. Roots of apparently healthy plants that are near plants showing early symptoms sometimes reveal the fungus in the first stages of its action. Whitish or tan threads of the fungus mycelium overrun the roots and break down the outer cells of the root, producing slightly discolored, depressed areas or lesions. Development of the lesions destroys the cortical tissues, and the organism penetrates into the woody central cylinder.

The mycelium assumes a tan or buff colour. Individual threads join together to form fuzzy strands. In more advanced stages, the cortical tissues completely disintegrate, and the infected part of the central cylinder becomes red or brown, quite different from the white to cream colour of the adjacent healthy tissues.

This fungus has several forms, or stages. The vegetative stage consists Of individual threads filaments, or hyphae of the fungus, or the threads may fuse to form mycelial strands. Filaments and strands may push through the soil until they reach another plant whose healthy roots they envelop and penetrate; failing that, they grow until their food supply is exhausted. Young filaments and strands are light-colored. Later they become cream to buff to brown.

A second stage is the spore mat, or fruiting, phase, which arises from the vegetative strands. One sometimes sees spore mats in cotton fields but more often in alfalfa fields. In warm weather when the soil is damp, the irregular- or circular mats appear on the surface of the soil. If conditions are to their liking, they may become a foot or more in diametre.

The mats develop rapidly. They appear first as a fluffy white fungal growth on the surface. They become creamy white or tan or buff in a few days. The mature mats are composed almost entirely of minute pores. Attempts to germinate the spores so far have been unsuccessful. The function of the spores is unknown. A third phase is the sclerotial, or resting, stage. Mature sclerotia are light brown to dark brown and as tiny as mustard seed.

In shape they are irregularly round. They are produced singly or in chains or clusters. Sclerotia develop on the mycelium or strands of the fungus, appearing first as slight, spindle-shaped swellings. They attain full sise in 4 or 5 days and mature in 10 to 14 days. The fungus produces them apparently in response to an abundant food supply. They have been found in quantities to a depth of 8 feet in the Texas blacklands.

The greatest concentration occurs at a depth of 1 to 2 feet. If moisture and temperature are right, the sclerotia germinate readily, and produce the typical Phymatotrichum mycelium and strands. Sclerotia may remain viable in the soil for at least 12 years, and upon germination can reinfect the roots of susceptible plants.. The fungus spreads mainly by the growth of the mycelial strands from root to root of the host plants or by free growth through the soil.. There seems to be little spread by ordinary tillage practices or on farm equipment.

A danger exists, however, of spread in the movement of nursery plants from an infested area, either as mycelium on the roots or as sclerotia in the ball of soil. Irrigation water or movement of surface water after moderate rains apparently does not spread the fungus, but heavy rains and gullying of the soil could possibly move sclerotia and set up new areas of infections.

Experiments with a number of field crops have established that rotations in which cotton is grown in alternate years are of little value in reducing losses from root rot. Rotations in which nonsusceptible crops occupy the land for 2 of 3 years give moderate protection. Marked and consistent control is obtained in rotations with nonsusceptible crops in which cotton is grown i year in 4. Because many of the tap-rooted weeds are susceptible to the fungus, a rigid weed-control programme is necessary during the rotation.

Other practices have value, although in themselves they are not considered to be economical or adequate control measures. Clean fallow for 2 years does not give consistently effective control in the cotton crop following the fallow period. After 8 years of clean fallow, the disease appears on the first crop of cotton and becomes increasingly severe on successive crops.

Deep tillage has given some evidence of control the next year, but the effect does not consistently persist beyond 1 year. Early-fall plowing of infested fields reduces somewhat the amount of disease in the following crop, but it is not cumulative in effect. A slightly better control is obtained by the combination of nitrogen fertiliser and early-fall plowing.

Inorganic fertilisers alone, particularly those high in nitrogen and phosphorus, increase the yield of cotton on infested soils, but their economic value has not been proved. Many soil disinfectants or fungicides have been tested against root rot, but the cost of materials or application precludes their use under ordinary field conditions. The search for varieties of cotton resistant to the disease has yielded negative results.

Early-maturing varieties of upland cotton, however, will set a partial crop of bolls before the period of greatest mortality, and their use is recommended on infested soils. Practical control of root rot in the irrigated Southwest has been obtained by the incorporation of large amounts of corral or other organic material into trenches or furrows.

The furrows are then covered over, beds are prepared above them, and the land is irrigated well before planting time. Yearly treatment of experimental areas in that way has reduced the disease greatly, although complete eradication is not attained. In nonirrigated places in the central blacklands of Texas, that principle has been applied successfully through the growing and turning under of legume crops in rotation with cotton.

Sweetclovers, such as Hubam and sourclover (Melilotus indica) as winter and winter-summer crops increase the nitrogen and organic content of the soils, reduce the amount of root rot, and increase the yield of cotton. In both irrigated and nonirrigated areas, control through the application of organic materials to the soil appears to be linked with the increased microbiological activity incidental to the decomposition of the organic matter.

Under laboratory conditions, both sclerotia and mycelium of the fungus are largely eliminated in organic-amended soils. In field experiments that used organic amendments (barnyard manure or sorghum fodder) and October plowing, the incidence of root rot was much less on the following crop of cotton, and no sclerotia were found in the upper 12 inches of soil. In adjacent plots that received only normal bedding operations, root rot was severe, and sclerotia occurred abundantly at the 6-inch to 12-inch level.

The importance of microbial action in the elimination of sclerotia and mycelium from organic-amended soils is strongly suggested by those data, but we cannot overlook the importance of the increased nitrogen content and the improved physical condition of the amended soils. In summary: The greatest promise of practical control of root rot appears to lie in the use of organic manure, particularly in the form of legumes.

Legumes may be used as winter green-manure crops and followed immediately by cotton, or they may be grown to maturity and the residues turned under during the summer. The use of organic manures, plus rotations, early-fall plowing, and early-maturing varieties, should provide satisfactory control of root rot.

VERTICILLIUM WILT OF COTTON

Verticillium wilt of cotton is caused by a soil-borne fungus. The disease occurs across the entire Cotton Belt from South Carolina to California. It is of major importance in the lower Mississippi Valley and the irrigated areas of the Southwest. Losses in yield from it may reach 50 per cent. Average losses of 10 to 15 per cent over large areas are not uncommon. Cool, wet weather favours the disease and plants may be attacked at any stage of development. The fibre from wilted plants is often damaged to the extent that the number of neps and manufacturing waste are increased, and the yarns are inferior in grade and looks.

The cotyledons of infected cotton plants become yellowish and quickly dry out. Young plants with three to five true leaves suffer considerable stunting. The leaves appear darker green than those of a normal plant and become somewhat crinkled between the veins. The amount of stunting apparently depends on the stage of development of the plant when it becomes infected. The outstanding symptom is the chlorotic areas on the leaf margins and between the principal veins, which make it look mottled.

In older plants the symptoms usually occur in the lower leaves first. They spread to the middle and upper leaves of the plant later in the season. The chlorotic areas gradually become larger and paler. Severely affected plants shed all the leaves and most of the bolls. Older plants may nevertheless survive the entire season and sometimes send up sprouts from the base of the plant.

Sometimes it is hard to tell verticillium wilt from fusarium wilt. In the early stages of verticillium wilt the vascular discoloration appears to be more evenly distributed through the stele of the plant at the ground line and to be lighter brown than the discoloration produced by fusarium wilt. In fusarium-affected plants, one or more leaves near the crown may wilt suddenly and die, while the other leaves remain apparently healthy.

That condition has never been found in verticillium wilt. In hop plantings in England, wilt appeared to spread in the direction of cultivation, since the spread of the disease was more rapid in cross-cultivated fields than in those cultivated in only one direction. Technicians at Shafter, Calif., found that the fungus spread more rapidly in the heavier soils. The spread usually was against the flow of irrigation water. In Mississippi, where the spread of the fungus was carefully studied, the fungus moved approximately 3 feet a year and spread out from centres of infection.

Diseased cotton stalks and leaf and stem trash from diseased stalks can carry the fungus over from one season to the next and spread the disease into new parts of a field.

Most of the commerical varieties of upland cotton (Gossypium hirsutum) appear to be susceptible to verticillium wilt. Varieties of Egyptian, Pima, sea-island, and some South American cottons (G. barbadense) have a high degree of resistance or tolerance. Numerous observations in the field indicate that the nutrition of the host has a pronounced effect on the development of verticillium wilt. Workers have found that it is particularly severe on soils high in organic matter and also that the disease is favoured by heavy application of organic matter to the soil. Efforts were made at Shafter to control the disease by applications of chemicals and soil amendments, but none of the practices appeared promising.

Despite the extreme susceptibility of most commercial upland varieties, some progress has been made through selection and breeding towards a high degree of tolerance or resistance. George J. Harrison at the United States Cotton Field Station, Shafter, has produced Acala 4-42, which has some tolerance to the disease. Further selections have been made that are even more tolerant.

A. R. Leding at the United States Cotton Field Station, State College, N. Mex., selected a wilt-tolerant variety Acala 1517 W. R. from Acala 1517. It was made available to growers in 1949. Of the cottons adapted to the Southeastern States, none has a high degree of tolerance to verticillium wilt. The fusarium-resistant varieties, such as Coker loo wilt, Coker 4-in-1, Empire, and Plains, however, have consistently yielded better than wilt-susceptible varieties when grown on soil infested with verticillium wilt.

At the Mississippi Agricultural Experiment Station, Hartsville was found to be highly resistant to verticillium wilt, but the variety was agronomically undesirable. Workers at several State agricultural experiment stations have used Hartsville in a selection and breeding programme designed to develop an agronomically acceptable variety that tolerates the wilt.

A grower should choose the most tolerant variety adapted to his particular location to be planted in fields infested with verticillium wilt. Any cultural practice that produces and maintains a higher soil temperature tends to reduce the amount of wilt. Mr. Harrison at Shafter, by carefully controlling soil moisture after the first of July reduced the incidence and severity of wilt. His method is based upon frequent light irrigations which permit more rapid warming of the soil following each irrigation.

ANTHRACNOSE AND SOME BLIGHTS

Cotton anthracnose, caused by Glomerella gossypii, is the primary cause of seedling blight, boll rot, and fibre deterioration in the more humid cotton-growing States of the South and Southeast. The boll rot phase was first discovered in Louisiana in 1890. The seedling blight and other phases of the disease were described in 1892 by George F. Atkinson at the Alabama Agricultural Experiment Station.

From that time until about 1920 the pink boll rot phase of the disease was extremely destructive. Anthracnose boll rot losses diminished with the coming of the boll weevil and the change to earlier, smaller, less vegetative varietal types.

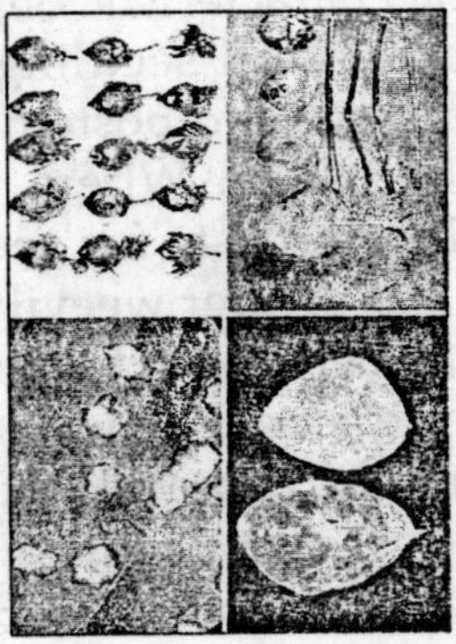

Upper left: Cotton anthracnose, here shown on bolls, reduces yields and discolors lint. Upper right: Symptoms of bacterial blight of cotton on stems, leaves, and bolls. Lower left: Enlarged lesions of the frogeye fungus on soybean. Lower right: Lower and upper surfaces of soybean leaves affected with bacterial blight.

The well-known sore shin of cotton seedlings is caused mostly by anthracnose. The fungus is mainly a secondary invader, entering through wounds, and is adapted to a semisaprophytic existence on injured or dead parts of the cotton plant. It is apparently universally present in all fields throughout the area of its distribution. Anthracnose and ascochyta blight have many features in common although caused by different pathogens.

Anthracnose exists in all humid cotton-growing areas. It has been distributed on seeds to all cotton-producing countries. In the United States it is widespread from Virginia to Texas and Oklahoma. It is delimited by the 40-inch rainfall line. The line extends north and south through eastern Texas and Oklahoma. West of that line, low rainfall and low humidity are unfavourable for the fungus.

The disease reaches its maximum intensity along the Atlantic and Gulf coasts and tapers off towards the inland areas in the Southeast, where somewhat lower humidity prevails. The fungus may be recovered in the laboratory from a high per centage of healthy-appearing leaves, stems, and bracts of most plants selected at random in fields from April to October particularly during periods of continued wet weather. Consequently the fungus is apparently present most of the time even on healthy-looking plant parts. No other disease of cotton is more widely prevalent. Anthracnose seedling blight losses were serious for many years before, when seed-treatment disinfectants came into general use. A 4-year survey of seedling diseases indicated that anthracnose was the predominant cause of seedling blights. The causal organism was recovered from 81.2 per cent of the diseased seedling samples. In many fields the characteristic blight lesions could be seen on the below-ground stems of all seedlings. Stand reductions are brought about by preemergence and postemergence blighting-off.

Skippy stands reduce yields. Where planting-over becomes necessary, there are additional costs for seed, labour, and reduced yields caused by lateness and greater weevil damage. Cotton planted over may yield only 60 to 80 per cent of the first planting. Losses from anthracnose boll rot were serious for many years after discovery of the disease in 1890. Commonly, damage was estimated from 10 to 70 per cent of the crop. Losses were much less after coming of the boll weevil and are now estimated at from 0.5 to 3 per cent, depending on weather conditions.

The anthracnose fungus is often associated with Alternaria, Fusarium,other fungi, and bacteria in boll rot lesions. A 4-year survey made by Paul R. Miller and Richard Weindling of the Department of Agriculture showed that the anthracnose fungus was present in 67.8 per cent of diseased boll samples. They considered it to be the major boll rot disease.

Losses are caused by a direct invasion of the unopened bolls with destruction of the seed and lint, by invasion of the partially open bolls causing hard locks, and by weakening and staining of the fibre. Symptoms on cotyledons are usually diseased areas on margins or small reddish or light-colored spots. Diseased seedlings show reddish-brown lesions below ground. The lesions may be on one side of the stem, or they may surround it and extend down on the root.

Many seedlings are killed before or after they emerge. Others survive with the change to more favourable growing conditions. With the falling of diseased cotyledons, anthracnose symptoms disappear over summer until lesions appear on bolls.

On the bolls the disease appears as small, round, water-soaked spots,which enlarge and become sunken and brownish in colour. A sticky mass of spores comes over the surface of lesions on bolls. The lesions may occur near the tip or at any other point. Often they are associated with wounds made by boll weevils. The bacterial blight pathogen often may be the primary invader, with anthracnose following in the same wound and often associated with other fungi. The lint and seed are rapidly invaded once the disease gets through the husk of the boll. P. B. Marsh and his associates, in studies in 1950 at the Pee Dee Agricultural Experiment Station, described "tight lock," caused by anthracnose and other fungi. Invasion of the lint and seed after cracking of the bur in wet weather produced a hardened, discolored lock. Long periods of rainy weather retard drying and fluffing of the lint and provide favourable conditions for development of fungi on the seed and lint.

Planting seed of cotton are almost always infested with the anthracnose fungus. H. W. Barre in studies at the South Carolina Agricultural Experiment Station found the fungus growing through the seed coat into the embryo. Seed fully invaded failed to germinate. Those that were partly invaded produced diseased seedlings. C. W. Edgerton, in studies at the Louisiana Agricultural Experiment Station, found by washing the seeds and counting the spores that as many as 8,000 spores were carried by each seed.

Paul R. Miller, in studies at the Clemson Agricultural Experiment Station, made other studies on spore load. Samples of cottonseed collected at random from gins in South Carolina in 1941 showed as many as 80,000 spores to a seed, although the average was much lower. Germination tests from the same lots of seed indicated that most of the seedlings had anthracnose infections even though few surface-borne spores were present.

The small amount of boll infections in many fields was not enough to account for the heavy spore loads carried by seed. Mr. Miller found that some of the spores came from the trash brought to the gin with the seed cotton. He also determined that clean cotton lots became contaminated with anthracnose conidia when ginned after infested lots. Thus the gin caused thorough contamination of all seed in an individual lot and served to contaminate several lots of clean seed ginned immediately following the ginning of infested cotton.

Rain drops splash spores from the soil onto the plants, from leaf to leaf, and from plant to plant. Infection is favoured by moderate temperature and high moisture. The fungi become more easily established in tissues affected by some kind of injury. There are no obvious symptoms of anthracnose infection in summer.

Essentially, the organism exists as a saprophyte in injured tissues. It may also live on honeydew on leaf surfaces. During dry periods it remains dormant and survives for long periods as mycelium in tissues or appressoria attached to the surface. With the return of wet periods it builds up to epidemic proportions and becomes destructive on the bolls. Infections on the boll gain entrance through the corolla, through bacterial blight lesions or insect wounds,through the sutures separating the loculi, or directly through the uninjured surface. The lint and seed of bolls beginning to open are invaded directly to produce tight lock without invasion going through the boll wall. Anthracnose seedling blight can be readily controlled by seed-treatment disinfectants. Beginning in 1936 uniform seed-treatment tests were conducted across the Cotton Belt by members of the cotton disease council of the southern section of the American Phytopathological Society.

An average increase of 30 per cent in emerged seedlings was obtained from application of dust disinfectants. After testing a large number of materials, the committee recommended the use of organic mercury compounds (5 per cent Ceresan and Ceresan M) and zinc pentachlorophenate (Dow 9 B). Those two volatile materials largely eliminate surface-borne anthracnose spores and give some protection to the young seedling in the soil. The council has also recommended the use of slurry treaters to replace the dust applicators. The slurry application largely eliminates the obnoxious dusts and reduces the danger to workers handling seed in processing plants.

Anthracnose boll rot, tight lock, and fibre deterioration still are serious problems. Losses are greatest in the Coastal Plain where cotton opens earlier and during the September storm period. Earlier recommendations for control of anthracnose boll rot included 1-year rotations to eliminate winter carryover of the organism on crop refuse and the selection of planting seed from healthy bolls or from noninfested fields.

It was found that the seed-borne fungus died out or was greatly reduced if the seed were stored for 18 months. Thus 2- or 3-year-old planting seed were recommended. Fall plowing was found beneficial for destruction of the stalks and other crop refuse before spring planting where rotations were not possible. Those measures, while effective, were not generally practical or widely adopted by growers. Defoliation can be the most effective method of reducing boll rot losses from anthracnose and other rots. Good defoliation is obtained when calcium cyanamide is applied as a dust at the rate of 30 pounds an acre. Removal of the leaves permits rapid drying of the bolls and lint after rains and dews and reduces boll and fibre losses.

Other practices of value in reducing boll damage are the planting of varieties with smaller leaves and open-type growth, limiting vegetative growth of the cotton plant by reducing nitrogen applications, controlling insects that damage bolls, and controlling grass and weeds that hinder air circulation. Varieties have different degrees of resistance to the pink boll rot phase of anthracnose. Somewhat resistant varieties are Toole, Dixie, Dillon, Express, Rowden, Cleveland, and Deltapine. Other varieties considered susceptible are Cook, Half and Half, Hi-

Bred, Lone Star, Wilds, Triumph, Trice, and Stoneville. The lack of recent severe infestations of anthracnose boll rot has prevented a classification for resistance of newly introduced varieties. The diminution in anthracnose boll rot may be attributed partly to the discontinued production of many of the older susceptible types in favour of the newer, less vegetative, earlier, resistant ones.

Resistance to anthracnose seedling blight has not been investigated thoroughly. As considerable success has been achieved in controlling seedling blights with dust disinfectants, an intensive effort has not been made to obtain resistant varieties. It is generally agreed that all commercial varieties are quite susceptible in the seedling stage. Ascochyta Blight, commonly called wet-weather blight or wet-weather canker, is the most sporadic of all cotton diseases. The causal organism is Ascochyta gossypii. The distribution of ascochyta blight parallels the distribution of anthracnose.

It occurs in all cotton States from Virginia westward to the 40-inch rainfall line in eastern Texas and Oklahoma. In 1950, a relatively wet year, it was reported prevalent in central Texas and as far west as Lubbock. The most serious occurrences of the disease have been in the Piedmont areas of the Carolinas and Georgia and in the northern parts of Alabama, Mississippi, and central Arkansas. Losses from the blight result primarily from reduction or loss of stands in the young plant stage. These losses were generally not recognised by pathologists until 1947.

From 1947 through 1950, a 4-year period of relatively heavier rainfall years, losses were common and serious in areas involving the northern parts of the Carolinas, Georgia, Alabama, and Mississippi. Serious losses occur during periods unfavourable for plant growth brought about by moist, cool weather. Plants 3 to 8 weeks old are particularly susceptible. The more spectacular but less damaging stem canker has occurred sporadically at infrequent intervals since 1914. Additional but minor losses are caused by leaf blighting and partial defoliation of older plants and boll rotting.

The earliest infections are produced primarily from spores that overwinter on the old plant refuse in the field and are splashed by rain or blown to the leaf surfaces. The spores also over-winter on planting seed that is not treated with a disinfectant. Earliest symptoms are small, circular white spots on cotyledons and leaves. The lesions enlarge, often coalesce, become brown and roughened, and often fall out, leaving a ragged appearance. Cotyledons and leaves often are defoliated and the young stems are left bare. Additional infections girdle the stems and kill the terminal buds and adjacent stem tissues;

the result is loss of stands. From June to August, conspicuous stem cankers occur at the branch axils and centre about the stipules. Cankers are from one-half inch to an inch long, dark brown, and ragged at the edges. Sometimes the stems or branches are killed by cankers, but usually stem cankers cause minor damage.

Stem cankers occur only during periods of several consecutive days of cloudy, wet weather. With the return of dry, hot weather, the disease is checked and canker symptoms are quickly obscured by new growth. During wet weather in July and August, older leaves frequently become infected. The lesions may spread quickly and consume most of the leaf area, so that there is extensive defoliation, particularly of the lower leaves. Boll rot lesions are rough, circular, and brownish. Invasion of the seed and lint complete destruction of the boll.

Excellent control of ascochyta seedling blight is had by seed treatment and crop rotation. Primary infections originating from seed-borne spores are largely eliminated by seed treatment with disinfectants. Rotation eliminates the overwintering inoculum that originates in the cotton plant refuse from the previous crop, and excellent control of seedling disease phases is obtained. Partial, but generally satisfactory, control is obtained by plowing old stalks and leaves under deeply in the fall and planting flat to leave such material buried. Other practices that reduce the amount of old plant refuse left on the surface of the soil at planting time may be beneficial.

We have little information on resistance in varieties. Studies made at the Georgia Agricultural Experiment Station in 1947 indicate Empire, Deltapine 15, and Stoneville 2B have slightly more resistance to stem canker in older plants than several other commercial varieties. No varieties have enough resistance to withstand the seedling blight attacks, which are the most destructive phase of the disease.

Bacterial Blight of cotton was first described in 1892 by George F. Atkinson in studies conducted at the Alabama Agricultural Experiment Station. He named the disease angular leaf spot and found a bacterium to be the causal agent. The thin-walled parenchyma cells of leaves, stems, and bolls are attacked. The stems may be girdled and show the "black arm" symptoms. Spread of the bacteria along the leaf veins is commonly called vein blight. Lesions on bolls are commonly called boll rot or boll blight. The common name bacterial blight has been adopted to include all those symptoms. The causal organism is Xanthomonas malvaceorum.

Bacterial blight is prevalent in all cotton-producing areas. It may have originated in India, the centre of origin of Old World cottons. It apparently has been associated with cotton in the United States from the beginning of cotton culture. The disease is worst in the subhumid and semiarid areas of the Cotton Belt where the rainfall varies from 10 to 30 inches. Epidemics occur almost every year in the High Plains of west Texas, in the Pecos and Rio Grande Valleys of New Mexico and Texas, and in Arizona and Oklahoma. The disease has been of minor importance in California. In the mid-South and Southeastern States the occurrence is universal, but damage is less serious and more sporadic. Winds of cyclonic intensity accompanied by rain sometimes cause epidemics of the disease over large areas in the Southeast.

Losses result from reduction in stand, defoliation of leaves, stem blighting, shedding of small bolls, boll rotting, and reduction in grade from lint staining. The American-Egyptian irrigated varieties of the Southwest may become complete crop failures during serious outbreaks. In 1949 the yield reduction was estimated at 35 to 50 per cent on 40,000 acres in New Mexico.

For upland varieties in the United States as a whole the damage is from 1 to 2 per cent of the crop. The southwestern upland crop, from Oklahoma westward to Arizona, is consistently affected to the extent that bacterial blight is considered the major disease problem. Besides lower yields largely due to leaf defoliation, the yellow bacterial slime stains the fibres and lowers the grade and price received by the grower. In the South and Southeastern States the blight bacteria provide primary lesions on bolls through which the boll-rotting Diplodia, Alternaria, Glomerella, Fusarium, and other fungi gain entrance and rot the seed and fibre.

Branches may be cankered or girdled and turn black. This symptom is known as black arm. It occurs commonly in the more susceptible American-Egyptian varieties. Occasionally the Acala upland varieties also show black arm symptoms. Spread of the organism along the midrib and major veins produces the commonly known vein blight. Younger leaves are more susceptible to vein blighting than are fully mature and older leaves.

Infections originating in the corolla may spread to the young boll and cause extensive shedding. On older bolls the lesions are circular and shiny and later become sunken and browned and blackened. Penetration of the blight organism through the husk wall permits staining of the fibre and invasion of the seed. Infected bolls become deformed and open prematurely.

Most of the infections of seedlings in spring are caused by bacteria carried over the winter on seed. The bacteria may be attached to the seed coat or seed fuzz or inside the seed coat. Seed harvested from heavily infested fields may produce seedlings with 25 per cent infection. Such early and heavy infestations often produce blight epidemics by the end of the growing season. The blight organism may also overwinter on plant refuse left in the field in arid areas. Volunteer seedlings from diseased bolls or old plants which survive the winter may carry the disease over and infect the new crop. Bacteria on infested seed are washed to the germinating embryo and initiate the new infections.

The primary lesions produce inoculum, which is splashed to the lower leaves to initiate secondary infections. The disease is carried through several generations on the leaves until the bolls are formed. Greatest damage by the disease is caused by boll infections. High temperatures and high humidity are favourable for maximum infection. The infrequent rains accompanied by strong winds of the High Plains area of the Southwest are most favourable for epidemic spread.

Relative scarcity of the disease in California is attributed to the higher elevation and average lower temperatures and scarcity of rainfall. Blight bacteria gain entrance to the host tissue through stomata and wounds. Entrance to leaf tissue is generally on the lower surface, where the stomatal openings are more abundant. Bacteria are transported by water to the stomatal cavity. The organism characteristically invades the parenchymatous tissue, made up of the thin-walled cells.

Less frequently they enter the xylem, or water-conducting vessels, of the leaf petiole, stem, or leaf vein. Once inside the stomata, the bacteria multiply and crush adjacent cells and continue to utilise and multiply on the destroyed cell contents. A bacterial exudate is produced and accumulates on the surface of the wound or drops off to other leaves or to the soil.

The exudate dries as a thin, yellowish, translucent film. This material is readily dispersed in rain water and serves as inoculum to further spread the disease. Water congestion of the leaf tissue caused by hard, blowing rains facilitates infection. Wounds made by insects also facilitate infections. Hailstorms which injure all parts of the plant may be followed by blight epiphytotics. The incubation period on bolls and leaves is 8 to 10 days, from time of inoculation to earliest appearance of symptoms.

The world-wide distribution of bacterial blight is attributed to the seed-borne bacteria. Spread in the field is largely attributed to wind-blown rain water and splashing from plant to plant and leaf to leaf by falling water. Bacteria are also carried by flowing of surface water. In irrigated fields the spread of the disease may be traced in the direction of the water flow. Experimentally, dissemination of the disease in fields can be correlated with the direction and velocity of the wind during periods of rainfall. The spread of bacteria by dust storms has been observed in Arizona. The wind-blown rains characteristic of the High Plains of Texas and the irrigated valleys of New Mexico and western Texas are more effective in spreading the disease than the rainfalls of the mid-South and Southeast. The movement of fallen diseased leaves by strong winds also spreads the disease.

Control measures for the bacterial blight disease have been directed towards eliminating the overwintering sources of inoculum on the seed and in the field. When F. M. Rolls found blight bacteria adhering to the seed fuzz and seed coat he demonstrated the possibility of removing them with sulfuric acid in experiments at the South Carolina Agricultural Experiment Station in 1915. With concentrated sulfuric acid all fibres are removed from the seed. Any adhering bacteria on the seed coat are also removed. Washing the acid off of the seed may remove any lightweight seed or floaters and give a high-grade planting seed.

Beginning about 1930, organic mercurial dust disinfectants became available for seed treatment, which also largely controlled seed-borne bacteria. A combination of delinting and dust treatment became widely used in areas where serious bacterial blight losses occurred. As an outgrowth of delinting efforts to control blight, commercial processing plants are now operating across the Cotton Belt and delinted seed of all major varieties is available. Many growers prefer the delinted seed for more precise planting with tractors.

Other practices for blight control are the selection of seed grown in disease-free fields and the destruction in early fall of the previous crop refuse by deep, clean plowing. Any volunteer seedlings are also destroyed in spring before planting.

Although a fair degree of success in controlling blight by sanitation, delinting, and seed treatment is obtained, the results are not always satisfactory. Internally borne infections often survive the treatments and initiate disease centres. Because all growers sometimes do not cooperate in the blight-control programme, the spread of infection from an adjoining farm may cause losses. Pathologists believe therefore that the only fully satisfactory possibility for control is the development of blight-resistant varieties.

The severe losses from bacterial blight in the Gezira of the Sudan caused British workers to make an intensive search for resistance to bacterial blight. The work was done by R. L. Knight and associates in studies conducted at the Agricultural Research Institute, Khartoum, Sudan. An upland-type cotton, Ugandi B31, was found to have a high degree of resistance. It has been used in breeding programmes. A blight-resistant upland, Stoneville 20, was discovered in 1939 by D. M. Simpson and Richard Weindling in breeding plots at the Tennessee Agricultural Experiment Station. Stoneville 20 is not a desirable commercial type, but it has no bad characteristics and is valuable as a source of resistance. In 1953 it was being used as a source of resistance in several breeding programmes in the Cotton Belt.

A satisfactory technique for artificial inoculation had to be developed before progress could be made in selecting resistant plants in the breeding programme. Mr. Knight at Khartoum learned that he could get uniform infections by soaking 10 pounds of diseased leaves in 40 gallons of water and using the liquid as a spray. The technique was improved by Richard Weindling in research work he did at the South Carolina Agricultural Experiment Station.

He found that a single potato-dextrose agar culture grown in a petri plate and diluted with 2 1/2 gallons of water provided a more satisfactory source of inoculum. He also discovered the best time to spray was from mid-morning until noon on sunny days when the stomata were wider open. The lower leaf surface due to the larger number of stomata also gave more infections than the upper surface. Additional improvements in the inoculation technique now indicate that by using an orchard sprayer, with 400 pounds pressure to the square inch, and forcing the bacterial suspension into the stomata very satisfactory infections are obtained on a field scale.

A technique for infecting greenhouse seedlings has also proved useful in the search for blight resistance within established varieties. The seed are first soaked in a bacterial suspension, planted in a greenhouse bench, and examined on emergence for sise of bacterial lesions. The lesions are nonexistent, or very small, on resistant plants. Resistance to leaf blight, black arm, and boll blight were all found to be positively correlated, and therefore the leaf symptoms alone could be used as an index to resistance within the individual plant in breeding work.

The inheritance of resistance to blight has been studied by Mr. Knight. In a survey of all cotton species and a large number of species types and varieties, he found five factors for resistance, B_1 to B_5. The factors varied in the amount of resistance they contributed and were generally additive or accumulative in contributing to the resistance of the plant. In segregating populations the resistance factors segregated as dominants or partial dominants. The resistance in Stoneville 20 is largely determined by a single gene. Resistance is recessive and susceptibility is dominant. In addition to the major gene in Stoneville

20, several minor genes contribute to the resistance with additive effects. The minor genes account for varying amounts of blight tolerance in commercial upland varieties. Deltapine has a high level of minor genes, Stoneville 2B somewhat less, and Acala a very low level. Resistance can be established in American uplands by the accumulation of several minor genes centreed about a major gene, as found in Stoneville 20.

The breeding of blight-resistant varieties adapted to all sections of the Cotton Belt is an extensive project and was well under way in 1953. L. S. Bird and L. M. Blank demonstrated the possibilities in breeding for resistance and outlined a backcross method in studies conducted at the Texas Agricultural Experiment Station. Four varieties Stoneville 2B, Deltapine, Empire, and Coker 100 wilt were each crossed with Stoneville 20.

Using the backcross method and the inoculation technique developed by Weindling, they were able to obtain satisfactory blight-resistant commercial types within 5 years. A search was started for naturally occurring resistant individual plants within a number of additional varieties. The work is done by growing several thousand plants in a field planting and inoculating them with a power sprayer. Promising plants have been found in several varieties. Pathologists are confident that all commercial varieties will sometime be bred resistant to bacterial blight.

BACTERIA AND FUNGI ON SEEDLINGS

Cotton seedlings are subject to attack by several fungus and bacterial diseases, especially when soil temperatures are low and wet weather prevails in the spring following planting. Some of the diseases are caused by strong pathogens, which deplete the stand to such an extent that replanting becomes necessary.

The additional expense for more seed and labour for replanting and a reduction in yield (which ranges as high as 15 per cent) because of late planting are considerable. Many species of fungi, bacteria, and nematodes have been encountered by pathologists in studies of seedling diseases of cotton, but little is known regarding the pathogenicity of some of them. I limit this report mainly to the pathogens whose distribution over the Cotton Belt and effects on seedlings are well known.

The name sore shin apparently was first used to describe the symptoms produced by damping-off of cotton seedlings in Alabama by George F. Atkinson in 1892. The disease is caused by the soil-inhabiting fungus, Rhizoctonia solani. It occurs throughout the Cotton

Belt and is considered to be the most serious seedling disease of cotton in Oklahoma, Texas, and the Western States. It is also frequently prevalent and destructive in the Mississippi terrace and delta soils of Arkansas, Louisiana, Mississippi, and western Tennessee and in other heavier soils of the middle South. It attacks cotton seedlings only under conditions favourable to the fungus and unfavourable to the plant, such as cold, moist weather. The disease is also more severe when seedlings are damaged by thrips.

When seedlings are attacked by the sore shin fungus, usually within 5 or 10 days after planting, dark- to reddish-brown cankers develop on the stems near the soil line. In severe attacks, the cankers encircle the stems or penetrate so deeply that the plants fall over and die. The attack usually occurs in early-planted cotton before the formation of true leaves, but the fungus may persist and continue to restrict root development and top growth until the plants begin to flower. Infection also occurs on the leaves late in the season in some States. Some defoliation results, but that is of little economic importance.

Tests have been made of various chemical and fungicidal materials for possible control of sore shin of cotton, but we had no satisfactory control measure in 1953. Certain practices will reduce losses.

First is the thorough preparation of the seedbed so that rows are well pulverised at the surface, firm beneath, and slightly raised-4 to 6 inches to give good drainage.

Second is the planting of seed that is certified as to viability and has been treated according to accepted practice for control of seed-borne diseases.

Third is to avoid planting too early. Early plantings usually are subjected to cool nights and moist weather which favour attack by the disease. With the newer and more effective insecticides now available for insect control, the planting date for cotton could be delayed with safety in many areas for 10 or 14 days or until the soil warms up.

Fourth is speeding up germination by using reginned or acid-delinted seed and liberal fertilisation so as to give the seedlings a vigorous start.

Other soil-borne organisms that attack cotton seedlings are Fusarium oxysporum, F. vasinfectum, the wilt fungus; Fusarium moniliforme and other species of Fusarium, which cause rots and blights; Thielaviopsis basicola, the cause of root rot; species of Pythium that cause damping-off; Sclerotium bataticola and S. rotfsii, which cause rots; and the parasitic nematodes Meloidogyne incognita var. acrita (which causes root knot), Pratylenchus leiocephalus (the meadow nematode), species of Trichodourus (which cause stubby root), and other nematodes, such as Helicotylenchus nannus and Rotylenchulus reniformis, which affect cotton and are known to be associated with the incidence of fusarium wilt.

Most of those organisms are pathogenic on seedlings and cause damage in some areas, but, with the exception of the root knot nematode, which frequently reduces seedling stands of Egyptian cotton in Arizona and New. Mexico, they are of minor importance in the complex of seedling diseases. The anthracnose fungus, Glomerella gossypii, is the main pathogen responsible for seedling blight and damping-off in all of the cotton-growing States east of the 40-inch rain belt of Oklahoma and Texas.

In a survey conducted from 1938 to 1941 by pathologists of the Department of Agriculture in 14 States, the anthracnose fungus was found to be the predominant organism affecting cotton seedlings throughout the Southwestern and Mississippi Valley States. Its distribution on seedlings also coincided with the area affected with anthracnose boll rot. It occurred in 81.2 per cent of seedling samples, but in Texas and Oklahoma the occurrence of the fungus was limited to the eastern parts of the States. The absence of the disease in the western belt is apparently due to dry conditions and high temperatures that prevent the survival of the fungus during the interval between the seedling blight stage and the boll rot stage.

Seed become contaminated at the gin with anthracnose spores, which are present in such trash as infected leaves, bracts, stems, and bolls. All such material may carry heavy loads of spores, which are mixed with the seed in ginning and adhere to the fuzz or linters. In studying the relation of germination to contamination of cotton seed by the anthracnose fungus, Richard Weindling and P. R. Miller, of the Department of Agriculture, find no relation between the sise of spore loads of Glomerella gossypii on a given sample of seed and the per centage of germination of the seed. They emphasise, however, that the sise of the spore load seems to influence the amount of post-emergence damping-off when the seeds are planted.

When anthracnose-infested seed are planted, the seed-borne spores become active and may attack the germinating seedling. Infected seedlings may be killed before emergence or after emergence or they may survive and overcome the disease. Affected seedlings have reddish or dark-brown lesions on the stems below the soil line and frequently on the roots. The cotyledons the seed leaves also may be attacked, the disease producing brownish spots that enlarge before the plant wilts and succumbs. If cool, moist weather prevails for several days after planting, infested seedlings continue to blight and damp-off to the extent that stands become skimpy and replanting is necessary. Data from 63 tests conducted between 1946 and 1948 show the value of seed treatment. The mean increase in surviving seedlings was 33, 30, and 26 per cent when fuzzy, reginned, and acid-delinted seed, respectively, were compared with untreated seed.

Of a large number of materials tested, higher per centages of emerged seedlings were obtained with the organic mercury dusts, ethyl mercury chloride, ethyl mercury phosphate, and ethyl mercury p-toluene sulfonanilide, which are sold under the trade names of 2 per cent Ceresan, New Improved Ceresan 5 per cent, and Ceresan M 7.7 per cent, respectively. Another material recommended by the committee is zinc trichlorophenate, 50 per cent, in a suitable diluent. It is marketed under the trade name Dow 9-B.

The materials are applied at the following rates: Fuzzy seed, 1 1/2 ounces to a bushel; acid-delinted seed, 2 ounces to 100 pounds; and reginned seed, 3 ounces to 100 pounds. They also are wettable and may be applied by the slurry method, which has given emergence values of seedlings comparable to the dust treatment. The slurry method involves the suspension of the disinfectant in water and its application to the seed in that form rather than as a dust or powder. The method eliminates flying dust during seed-treating operations, ventilation or dust exhaustion, and the use of masks by workmen. Another advantage of the slurry method over dusts is that it gives greater accuracy and uniformity of dosage.

Chemicals for treating seed may be applied as dusts or slurry in homemade rotary drums equipped with tight-fitting lids or in specially constructed motor-driven machines with treating capacities of 4 to 6 tons of seed an hour. Reginning, or machine delinting of cotton seed, is done extensively in the Central and Southeastern States. The linters are removed in the process, so that planting is more uniform, hill dropping equipment can be used, and seed and chemical dusts are saved.

Light reginning is preferable; close reginning may damage some seed. During certain years of the regional tests, reginned treated seed gave a higher emergence of seedlings than fuzzy treated seed, but in other years the differences were small. In 1938, for instance, reginned, Ceresan-treated seed in 21 plantings gave a higher mean number of surviving seedlings than comparable fuzzy treated seed. In 1939 in 18 regional plantings, the increase in number of surviving seedlings from reginned seed treated with 5 per cent Ceresan was much higher than fuzzy seed similarly treated. In other tests later, the number of surviving seedlings for reginned was highest in 1946, slightly higher for fuzzy in 1947, and practically the same for both seed sublots in 1948. Acid delinting of cotton seed is used more extensively in the drier areas of Texas and Oklahoma and in the irrigated areas of the western cotton States.

It is used to a lesser extent in the Mississippi Valley and the Southeastern States. Either sulfuric acid or hydrochloric acid gas is used for delinting. With the sulfuric acid method, the seed is mixed with the acid in delinting machines, which do the work rapidly and efficiently. After the linters are dissolved, the seed is thoroughly washed in large vats to remove the acid.

The lighter seeds, or floaters, are removed during washing. The heavier seeds, or sinkers, are given a final rinsing, usually in dilute limewater, dried, and retained for planting. With hydrochloric gas, the linters are removed by the gas from the heated acid. It has the advantage over the sulfuric acid method in that the seed is kept dry throughout the treatment.

The lighter seeds are removed with fans. Acid-delinted seed usually give a higher emergence of seedlings than reginned or fuzzy seed if weather conditions are favourable. But we have instances in which seed so treated gave only poor to fair germination during cool, wet weather. In 13 regional plantings in 1946, the surviving seedlings at final count (as per centage of seeds planted) were 73.0 for delinted, 61.7 for reginned, and 60.0 for fuzzy seed of the same lot. While acid-delinted seed gives good stands and requires less seed to the acre, especially if hill-dropped, the method is more expensive than machine delinting.

Bacterial blight causes death of seedlings and loss of young plants because infected stems are weakened by blight and break during wind storms. The disease occurs throughout the Cotton Belt. It is most destructive in the Southwestern States. First, small, round, water-soaked lesions develop on the cotyledons as they emerge from the seed coat. The lesions furnish inoculum for the developing true leaves, the bacteria later infecting the terminal bud and main stem. Delinting cotton seed with sulfuric acid plus fungicidal treatment eliminates blight bacteria from the seed.

THE LEAF SPOTS OF COTTON PLANTS

One or more of the leaf spots of cotton occur in practically every cotton-field when the crop is growing. Some have to do with maturity of the plant and have no great effect on its growth or yield. Others develop on seedlings or young plants and damage the normal growth and fruiting processes. The leaf spot organisms commonly cause shedding of the affected leaves, but some may injure the branches and bolls. This article deals mostly with the effect of some of the leaf-spotting organisms on the foliage of the plant.

The leaf spot caused by Alternaria tennis probably is one of the most common. We have known about it since 1918, when investigators learned that it was often linked with angular leaf spot, or bacterial blight, and that it was found on leaves throughout the growing season. Often it was found overgrowing old angular leaf spot lesions or in places where spider mites were doing harm. We have learned that it can infect undamaged leaves and may be considered as a weak parasite. Sometimes Alternaria may cause severe shedding of leaves.

In the early stages, the spots on the leaves have a pale-green area with an indefinite and irregular margin. As the area enlarges, the colour of the older part becomes straw yellow and then rusty brown. By then the spot contains a number of irregular concentric zones in the roughly circular lesions and the fungal spores may be found on the surface of the lesions.

We do not know exactly which environmental conditions favour the development of this leaf spot, but it has been reported on plants suffering from low vigour because of drought or deficiency of potash. Experiments indicate that plants can be inoculated most easily if humidity is high. Although the alternaria leaf spot is common, it is considered of less importance than other leaf-spotting organisms. Little therefore has been done on control measures.

Ascochyta Leaf Spot, canker or blight, wet-weather canker, and wet-weather blight are names commonly given to the disease caused by the fungus Ascochyta gossypii. As the names indicate, the disease may manifest itself as a seedling blight, a leaf spot, a stem or petiole canker, and as a boll spot. Ordinarily several phases of the disease may be seen on a plant. The first symptoms are small, round, white, purple-ringed spots on the cotyledons and lower leaves. The spots become somewhat elongated and raised on the upper surface.

Later they change to a light brown, the purple ring around the outside disappears, and the diseased tissue often falls out. If infections are numerous, irregular diseased areas result and are followed by drying and shedding of infected cotyledons and lower leaves. Pycnidia fruiting bodies of the fungus are produced irregularly or in concentric rings and are visible on the upper surface of the lesion. Growth of the plant is stunted. The upper small leaves, petioles, and buds are often infected, and the plant dies. The almost bare stems, with a few small leaves at the tip, are characteristic of the disease in the later stages.

Outbreaks have been associated with long periods of rain and cool weather. Ascochyta may be seed-borne, but the carry-over of diseased stalks and branches from the preceding year's crop may be of greater importance in setting up new infections. The disease can be spread in infected plant debris; local spread appears to be mainly by the flow of surface water from contaminated fields to lower places.

Rotation with other crops helps control it by eliminating the carryover of the fungus from one season to another. If rotation is not practicable, old cotton stalks should be plowed as soon as possible after harvest to obtain maximum decay or deep-plowed in the spring to reduce the amount of infected plant material on the surface at time of planting.

Seed treatment reduces the amount of carry-over on the seed and improves the initial stand but does not give protection beyond the early seedling stage. The Bacterial Blight disease, caused by Xanthomonas malvacearum, can affect all above-ground parts of the cotton plant. It occurs in all cotton areas. In the United States it is particularly severe in the Southwest.

It shows up on the leaves as angular, water-soaked lesions, which turn brown or black when dry. On the bolls, it appears as round water-soaked lesions, which are sunken and black when dry. On the stems and fruiting branches, it produces black, elongated lesions. Therefore the disease is commonly known as angular leaf spot, boll blight, and black arm.

On upland varieties it may cause a blight of seedlings, defoliation of plants, or shedding and rotting of the bolls. Upland varieties in the Cotton Belt are affected by the black arm phase, but it is more common on the American, Egyptian, and the sea-island varieties, which often show almost complete destruction of bolls and branches. Severe leaf infection in all varieties means extreme defoliation; slight to severe losses in yield of seed cotton may follow.

The bacteria that cause the disease may overwinter on the surface of the seed, within the seed coat, and on diseased cotton stalks and bolls from the previous crop. Volunteer seedlings from infected bolls can be responsible for early appearance and spread of the disease, particularly when splashing rain or irrigation water aid in the movement of the bacteria. The first evidence of the disease is seen on the under surface of the cotyledons as water-soaked spots, circular or irregular but without the marked angular appearance noted on the true leaves. Such lesions are first evident along the margins of the cotyledons and then spread inwardly. Later the part near the original site of infection becomes brown and dry, frequently distorting the shape of the cotyledon. The bacteria may move along the petiole of the cotyledon and into the stem of the young seedling, producing the typical water-soaked appearance of the tissues. Continued movement of the bacteria into the terminal bud region results in tissue collapse and death of the young plant.

On the true leaves the external symptoms appear as water-soaked lesions, first on the under surface and later on the upper surface. The lesions are bounded by small veinlets of the leaf hence their characteristic angular appearance. If the young leaves are held towards the light, the spots appear translucent. As the lesions become older, large amounts of bacterial slime are exuded and form a dry film on the discolored lesions, usually on the under side of the leaf. The infected area of the leaves finally becomes dry and sunken and turns reddish brown or black; often the nearby healthy areas become yellowish. At times the blight lesions along the veins of the leaf produce a symptom known as vein blight.

The bacteria are spread from older to newly formed leaves and eventually to the bolls by wind-driven and splashing rain. Dry, hot weather checks the progress of the disease. Control measures have centreed around the use of disease-free seed, attained through acid-delinting and the application of a chemical disinfectant. That practice has been successful, particularly in the rain-belt part of the production area. In drier parts of the Southwest, however, there may be more overwintering of the organism in undecomposed, infected plant material, and the use of acid-delinted and dust-treated seed has not been uniformly successful.

In order to reduce the carry-over of the blight organism in plant debris, we recommend rotation with other crops or the early and complete turning under of infected material, followed by irrigation to promote decomposition. But the use of resistant varieties is the only adequate control throughout the entire period of plant development. D. M. Simpson, at the Cotton Field Station, Knoxville, Tenn., found a high degree of resistance in one of his breeding strains of cotton.

It was released to other breeders as Stoneville 20 and has been used widely as the resistant parent in crosses with a number of desirable, but susceptible, commercial varieties. The qualities of the commercial parent are then regained through a series of back-crosses to this parent, accompanied by selection for the resistance and for good fibre properties and high yield. Field tests in 1950, 1951, and 1952 in Texas and New Mexico confirmed the practicability of combining blight resistance with the desired fibre and yield properties of the recurrent commercial varieties. Leaf Spot Due To The Attack of the Cercospora fungus (Cercospora gossypina) is common in almost every cotton field towards the end of the season. The spots it causes are rarely more than one-fourth of an inch in diametre and are roundish or irregular. The lesions have purple borders and white centres, which finally fall out and leave a riddled appearance. The fungus may attack uninjured tissues, but it seldom causes much defoliation and is of minor economic importance.

Ramularia Leaf Spot, also called areolate mildew and frosty blight, was first discovered in 1889 on leaves collected near Auburn, Ala. It has been found since in many Southeastern States and abroad. The disease usually occurs towards the end of the growing season. It may cause some leaf shedding, but it is not of great economic importance.

A typical symptom is the whitish growth of the fungus on the under side of leaves, which produces a frosty or mildew appearance. The spots are angular and bordered by the veinlets of the leaf. Viewed from above, the lesions are bright green to yellowish green, and only occasionally show the white coating so typical on the under surface of the infected leaves. Similar lesions may form on the bracts surrounding the bolls.

The Fungus that causes sore shin and damping-off of cotton seedlings, Rhizoctonia solani, has been reported also as a leaf-spotting fungus in Louisiana. In the early stage, light-brown, irregular spots of varying sise appear between the veins and are bordered by dark, purplish rings.

As the fungus advances, it causes the tissue immediately surrounding the spots to become chlorotic and the dead tissue in the centre of the old spot cracks or falls out. On the lower surface of the leaves the smaller spots may be covered by light- to yellowish-brown growth of the fungus. It might affect considerable leaf area, but it is not of serious economic importance as a leaf-spotting organism.

NONPARASITIC DISORDERS OF COTTON

Potash hunger of cotton was known as rust or black rust before 1892, when George F. Atkinson demonstrated that it could be controlled by additions of potash to the soil. He called the disease yellow leaf blight, but the name rust has remained in common usage. Scientists frequently use the term "rust or potash hunger" because it describes both the symptoms and the cause.

Rust occurs commonly on the lighter soils of the Coastal Plain areas of the Cotton Belt and in many other soils of several mid-South States, including Alabama, Arkansas, Louisiana, and Mississippi. W. W. Gilbert estimated the average damage for the entire Cotton Belt at 4 or

5 per cent in 1920, but greater use of balanced fertilisers in the affected areas undoubtedly has reduced the loss in recent years. Affected plants are usually stunted and fail to develop a normal green colour. The lower or older leaves are first to show the typical mottled appearance; the areas between the veins turn yellow. Towards midseason the symptoms are visible over the entire plant. The yellowish spots on the leaves enlarge and become reddish brown or bronzed.

Later the entire leaf may become blackened, curled downward, and ragged. Brown or black circular or irregular spots are often produced on many affected leaves by the growth of Alternaria and perhaps other secondary invaders. Leaves usually are shed prematurely and often the stalk is left bare. The number and the sise of bolls are reduced. Yields are severely lowered. Affected bolls do not flare open properly, and so the cotton is hard to pick. Lint and seed from affected plants are inferior.

The primary cause of rust is insufficient potash to meet the nutritional requirements of the plant. The condition usually is found on potash-deficient soils, but not necessarily so. Excessive applications of nitrogen or phosphorus, or both, often may accentuate severity. Over-liming or a scarcity of sodium, which substitutes to some extent for potash, likewise tend to increase rust. Lack of humus in the soil and improper drainage have also been listed as conditions responsible for or contributing to severity of the disease.

The application of fertilisers that contain ample potash to furnish the needs of the plant throughout the season is considered the standard method of control. The exact amount needed to correct the condition depends on the supply of other soil nutrients. A common practice on lighter soils, where early rains can easily leach out some of the available potash supplied before planting, is to apply some potash along with nitrogen as a side dressing after the plants are established.

Potash hunger should not be confused with the true rust of cotton, which is caused by attack of the parasitic fungus Puccinia stakmanii, and occurs in limited areas of the Western States. Fusarium wilt and root knot nematodes also are often prevalent in the soils where potash hunger is common. It is not always easy to determine the exact degree of damage associated with each cause in this complex.

The Crazy Top disorder of cotton was first recognised by C. J. King and H. F. Loomis in a field of Pima cotton near Scottsdale, Ariz., in 1919. Three years later it was observed in a field of upland cotton near Casa Grande, Ariz. By 1924 the disorder had become common in Arizona and by 1936 it was still considered one of the most important problems of cotton production in certain districts of the State. Mild stages of the disease were also known to have appeared in the San Joaquin and Imperial Valleys of California between 1926 and 1933. O. F. Cook was first to present an accurate description of the disorder. He named the disease acromania, but the term crazy top, which is descriptive of the abnormal branching and fruiting in the upper part of affected plants, has remained the one in common use.

The first symptom is an abrupt change in the type of new growth made in the top of the cotton plants. Fruiting branches frequently are replaced by vegetative branches, which have regrown in an upright position and are developing an abnormal or crazy appearance in the tops of the plants. Typical features of the abnormality include reductions and distortions of leaves, internodes, bracts, and floral parts. The leaves are small, rounded, cupped, and thickened.

Flowers are small and distorted. Sterility is common and may be expressed either as complete suppression of floral buds or as profuse shedding of flowers. The few bolls that develop are usually small and malformed and contain a reduced number of seed. The disease occurs only on calcareous soils and has been found almost entirely related to irrigation practice.

It is associated with the checking of growth from water shortages and the resumption of growth when abundant moisture is restored. The disease is controlled easily by applying irrigation water frequently enough to prohibit a checking of plant growth during the summer months. Water shortage may not be the only factor involved in bringing about the abnormalities maintaining organic matter at higher levels by rotation with alfalfa is beneficial.

Crinkle Leaf is the name given by D. C. Neal in 1937 to a peculiar disorder of cotton plants occurring in Lintonia and Oliver silt loam soils of Louisiana and Arkansas. The typical symptoms are a puckering, mottling, partial chlorosis, and distortion of the leaves. The branches are often fasciated and the floral buds, flowers, and bolls are distorted. The fibre from affected bolls is often so weak as to be considered worthless. Dr. Neal and H. C. Lovett demonstrated by experiments they conducted in Louisiana in 1937 and 1938 that the disease was associated with high soil acidity, calcium deficiency, and manganese toxicity.

The typical symptoms were readily produced by adding increasing amounts of manganese sulfate to pot cultures of cotton. But it was easy to control the disease in the field by adding limestone. At the higher pH produced by treating the soil with limestone, or with other material containing basic carbonates, the manganese apparently is precipitated, and so the plants remain healthy. Thrips Injury is not strictly a nonparasitic disease of cotton, but I include a brief description because its symptoms are not unlike those of certain stages of crinkle leaf and crazy top. Infestation by thrips frequently causes serious injury to cotton plants. Reduced stands or reduced yields result from stunting and delayed fruiting of the plants. Damage is severe only in the early-season development of the plants and is usually restricted to mutilation of seedling leaves as they unfold from the bud.

Often, however, the terminal bud is destroyed, so that the plant gets a stunted appearance. Until it outgrows the disorder around mid-season, the branching is bunched or fasciated. The leaves may appear torn, perforated, cupped, and distorted. Usually there will be little if any change from normal green colour. A serious aspect of the disease is that the injury and decreased vigour tend to make plants more susceptible to the sore shin disease of cotton seedlings caused by Rhizoctoma solani a fact demonstrated in experiments conducted by Neal and L. D. Newsome in Louisiana in 1950.

Considerable variation in susceptibility among commercial varieties of cotton was observed by W. W. Ballard in Georgia. He pointed out that dense pubescence of juvenile terminal leaves appeared to be associated with resistance to thrips in some cases. He noted also that the degree of injury was usually correlated with the amount of aborted terminal buds, although a few varieties exhibited a degree of injury independent of the proportion of terminal buds aborted. 2,4-D (2,4-dichlorophenoxyacetic acid), a hormone-type herbicide commonly used to kill broadleaf pests in lawns, is extremely toxic to cotton plants. The minute quantities that light on cotton plants when the drift from adjacent applications is not controlled can cause severe injury.

The danger has been thoroughly recognised since this weed killer was first introduced commercially in 1946 but, despite adequate and repeated warnings, cotton may be found injured occasionally from drift, from accidental contamination of other chemicals used on cotton, or from improperly cleaned machinery and containers that have been used with 2,4-D. The symptoms on cotton are characteristic and specific formative effects, except where the dosage is sufficiently high to kill the plants or plant parts outright.

The effects are first noted near the growing points and the characteristic symptoms are usually the result of an abnormal stimulation of growth of the affected parts. Leaves are greatly modified. They become narrow, closely and conspicuously veined, and deeply lobed. Affected flowers are modified in a manner similar to the leaf elongated and narrow. Bracts are modified. They become deeply lobed and elongated and they fail to separate; they grow as a sheath around the developing boll. When enough material is absorbed to kill or inhibit the terminal bud, the plant may become branched and fasciated, with slender and malformed leaves, branches, buds, and flowers. Affected bolls and squares may turn yellow and then drop, although some malformed and partially fertile mature bolls have been observed. Often, however, the squares or bolls die without shedding.

Excessive shedding of leaves as a result of 2,4-D injury has not been reported. If the tops of plants are damaged, they may send out strong laterals, which produce apparently normal branches. The fully grown leaves, branches, or bolls seldom show much injury from the low dosages.

The formative effects in cotton continue to develop for some weeks following application of 2,4-D but the persistence of the stimulus depends on the stage of plant development and the concentration applied. If young bolls are present, the seed may be, injured. The likelihood of damage is greater for seed developing from plants treated at the time of flowering and, normally, no effects on the seed will be noted if parent plants are treated before the young square stage of growth.

Some instances have been reported where applications of organic insecticides have been associated with injury to cotton plants similar to that produced by 2,4-D. Many are undoubtedly cases of contamination, but with the use of certain of the organic phosphates it may be the insecticide itself causing the injury. Wayne J. McIlrath has shown a striking similarity of symptoms produced by treating comparable plants with 2,4-D and with commercial HETP (hexaethyl tetra-phosphate), and to some extent with TEPP (tetraethyl pyrophosphate).

Chapter 5

Food and Feed Grains

The root rots of wheat, oats, barley, and rye are among the least conspicuous diseases of cereals, but they are among the most destructive. They are caused by many species of fungi, which are widespread and live on or in seed, soil, and dead plant refuse. They attack a large number of hosts and thrive under a wide range of conditions.

To the many types of root rots in cereal crops many names have been given seedling blight, take-all, root rot, basal stem rot, foot rot, snow mold, Victoria blight, the helminthosporium blight, stem break. I use the term root rot here to include all diseases that affect roots and basal portions of the culms, although occasionally I refer to special types of root rot. Root rots of cereals were prevalent in the United States many years ago, but they did not attract attention until about 1900. H. L. Bolley in North Dakota in 1909 emphasised the importance of root rots on cereals in the United States.

He proved that the root rots of wheat were caused by several fungi that accumulated in the soil, particularly if crop rotation was not practiced. He attributed the unproductiveness of certain wheat lands in the Northwest to root rots, instead of depleted soil fertility or to toxins. The discovery of wheat rosette (a virus disease) in Illinois in 1919, originally attributed to root rot fungi, and the severe outbreak of root rots in the rust nursery at St. Paul, Minn., in 1919 and 1920 stimulated further investigations.

In the United States the prevalence and severity of root diseases depends largely on the particular kinds of crop and varieties. In general, take-all is the most important root disease in winter wheat areas, cercosporella root rot is prevalent only in the Columbia Basin of the Pacific Northwest, and helminthosporium and fusarium root rots are usually more destructive in the spring wheat region. The diseases are not restricted to any region, however. For instance, root rot of oats, caused by Helminthosporium victoriae, has been destructive in all districts where varieties from Victoria crosses are grown. Root rots probably are present every year in all fields of cereals in the United States. Sometimes they cause enormous losses to wheat, oats, and barley. In some places they are the limiting factor in crop production and prevent the growing of certain varieties of cereals in a region, such as barley varieties susceptible to common root rot in Minnesota and oat varieties susceptible to helminthosporium blight in the United States.

The amount of damage they cause varies greatly from season to season and from one field to another and often is difficult to determine because of the nature of the diseases. Root rots usually are not conspicuous, like the rusts and smuts. Infected plants commonly reach maturity, although various organisms may injure their roots. Unless damage is severe and plants are actually killed in definite places, as frequently happens in fields infected by the take-all and snow mold organisms, damage caused by root rots frequently is overlooked. Infected plants yield less.

Their seeds are fewer, smaller, sometimes shriveled, and of poorer quality than those from healthy plants. Take-all is a major disease of wheat in Australia. Losses may be as much as 60 per cent. Heavy losses also have been reported in the United States and Europe. Two destructive epidemics of root rots of barley occurred in 1943-1944 in the upper Mississippi Valley. Barley production in Minnesota dropped from about 50 million bushels to less than 13 million. Yields dropped more than 30 per cent.

Helminthosporium blight of oats was first observed in Iowa in 1944. By 1946 it was prevalent in every important oats-growing State, and ruined many millions of bushels of oats. In Iowa and Illinois it caused an estimated loss of at least 20 per cent of the potential oats crop more than 20 million bushels in Iowa alone. The disease was so destructive in 1947 that the growing of varieties such as Vicland and Tama that were released to growers because of their resistance to crown and stem rusts actually had to be discontinued.

The Symptoms of some root rots, such as typhula and take-all, can be diagnosed readily. But it is nearly impossible to distinguish the symptoms of root rots caused by some organisms, and one must use laboratory procedures to identify the cause. The difficulty arises because the symptoms caused by different pathogens often overlap. They may be almost identical under some conditions.

Two or more pathogens may be associated on a given host. Saprophytes maybe present and may modify the symptoms. Root rots generally are characterised by seedling blight, stunting of plants, yellowing and bleaching of foliage, discoloration of roots and bases of the stems, and premature killing of adult plants. The same root-rotting organisms also may attack the upper part of the plants and cause head and seed blight, foliage lesions, and rotting of nodes. Take-all, caused by Ophiobolus graminis, is a common root rot of wheat. It also attacks rye, barley, and oats. The diseased plants frequently are stunted, ripen prematurely, and occur in patches, which stand out among the normal plants. The black lesions of the rotted tissue, sometimes with conspicuous mycelium at the base of culms, are easy to spot.

The root rot that is caused by cercosporella herpotrichoides is called foot rot or stem break. It makes eyespot lesions at the base of culms. Infected plants have a tendency to fall over as the heads fill and make patches of lodged and tangled plants. The pathogen attacks wheat, barley, and rye, but not oats. It may be the same as Leptosphaeria herpotrichoides, the cause of the destructive stem break of winter wheat in northern Europe.

The root rot symptoms caused by species of Pythium vary and are influenced largely by the crop and the organism: In Japan they are associated with certain types of snow mold on wheat; in Canada the disease is called "browning" because of the brown colour that follows the gradual dying of foliage; in the United States Pythium causes root rot of mature plants and seedling blight of wheat, oats, and barley. The symptoms produced by Pythium may be confused with other types of root rot. The leaves of infected plants, particularly the lower ones, tend to lose their green colour. Infected roots develop reddish-brown lesions. Snow mold is common in winter cereals in northern Europe. It is caused by several fungi, among them Calonectria graminicola, Typhula itoana, and Pythium species.

The outstanding symptom is a conspicuous mycelial growth, which covers the affected seedling as the snow disappears in spring. The root rot caused by Colletotrichum graminicola is an important disease on oats. The fungus produces a distinct black stromatic formation (cushion-like) at the base of the stem near the surface of the soil. The so-called common root rot fungi on wheat, oats, barley, and rye produce few distinguishing symptoms. They generally attack all underground Parts of the plant and cause varying degrees of rots on roots, crown, and basal parts of stems.

The diseased tissue often has large light-brown to black areas. Common root rot may kill plants at any time from germination until maturity. They may or may not be stunted. Seedling blight commonly occurs. Symptoms usually are not apparent on adult plants unless their bases are examined. Common root rot is caused by many species of fungi. Helminthosporium sativum, H. victoriae, H. avenge, Fusarium graminearum, F. culmorum, F. nivale, and Rhizoctonia solani are the more common ones. Root-rotting Fungi usually cause most damage when conditions are unfavourable for the development of the host plants. Temperature, moisture, the amount of inoculum, and mechanical injuries of the host, such as insect damage, have a bearing on the severity of root rots.

Rusts and foliage diseases may predispose plants to root rots. Sometimes that effect is indirect severe rust in the fall on winter grain may make plants more subject to winter injury, which, in turn, predisposes the plants to root rot. Soil temperature affects both host and pathogen, but its effect apparently is more important on the host. High temperatures, which are detrimental for normal development of wheat, oats, barley, and rye, favour the growth of many of the root-rotting fungi. Seedling blights and root rots generally are most severe at relatively high temperatures.

Early seeding of spring-sown grain therefore reduces the incidence of root rot and increases yield. The organisms associated with snow mold, however, are most destructive at relatively cool temperatures and cause the greatest damage while the hosts are covered with snow. Winter injury to fall-sown grain may predispose plants to root rots. Diseased seeds often are a source of primary root infection and are important in relation to seedling blight. Cereal seeds grown in the more humid parts of the United States are commonly attacked by a large group of fungi and bacteria.

Fungi of more than 25 genera and several kinds of bacteria have been isolated from wheat and barley kernels. Many are saprophytic or weakly pathogenic. Species of Alternaria, which usually are non-pathogenic, are the most common organisms associated with kernels of grain, but Helminthosporium species and Fusarium species are the most common virulent root-rotting pathogens prevalent in the seed. Preemergence killing, seedling blight, root rot, basal stem rot, and lowered yields frequently result when such seed is sown.

Striking differences exist in the prevalence of seed-borne parasitic fungi, especially Helminthosporium and Fusarium, not only in different seasons but in different localities in the same year. Seed of wheat, oats, and barley with 10 to 25 per cent infection with Fusarium and Helminthosporium are fairly common some seasons. Seed lots with more than 5o per cent infection are not infrequent. The per centage of seed infected with Helminthosporium species and Fusarium species is closely associated with the per centage of germination, stand, amount of seedling blight, and the number of stunted and deformed plants. Various kinds of mechanical injuries, such as cracking and chipping of the seed coat during threshing, may be the indirect cause of poor stands from seeds that are relatively free from root-rotting organisms. The injuries may not be enough to reduce germination in laboratory tests. Sometimes the cracks are so small that they can be seen only through a magnifying glass. When such seed is planted, especially under conditions unfavourable for germination, however, soil-inhabiting fungi may enter through the breaks and rot the seed before it can germinate. Treatment with a suitable disinfectant places a layer of protective fungicide over the openings and prevents the entrance of root-rotting organisms. The more extensive the seed-coat injuries, however, the greater the need for seed treatment and the greater the benefits from it.

Many Species Of Insects and nematodes attack the roots and basal portion of the stems of cereals and grasses. The bluegrass billbug, Calendra parvula, has received the greatest attention of technicians. It attacks the roots, crown, and lower internodes of the stem, often weakening the plants and providing avenues of entrance for rot-inducing fungi and bacteria.

As the external symptoms of insect injury are seldom conspicuous, the relationship of insects to the rot is apt to be overlooked. Decay usually starts around the wounds the insect makes when it feeds and deposits its eggs. It is most common on wheat, but it also attacks barley, rye, and many grasses. In 1939 the amount of infested wheat plants in the hard red spring wheat region ranged from a trace to 96 per cent. In southern Minnesota and eastern South Dakota, fields with 25 to 50 per cent infestation were not uncommon.

Studies made in the north central region of the United States by Earle W. Hanson, of the Department of Agriculture in cooperation with the Minnesota Agricultural Experiment Station, indicate that the greater the infestation of billbugs the greater the amount of root rot. Obviously symptoms of root rot do not necessarily indicate insect infestation, since root rots may develop in the absence of insects, but invariably the insect-infested plants are more severely rotted than the noninfested plants. Furthermore, varieties somewhat resistant to root rots may develop severe infection when injured by insects.

Some of the root-rotting organisms are relatively weak parasites. They can cause damage only when the plants are weakened or when entrance to the plant is provided through wounds, such as those caused by insects, nematodes, and mechanical injuries. The frass deposit within insect tunnels in the base of the plant makes a good medium for growth and multiplication of many fungi. The movement of the insect inside the plant helps to distribute the fungus organisms within the host. The injured tissues are readily attacked even by semi-weak parasites. Therefore rots that follow insect injuries usually involve much more tissue than that destroyed by the insect. Insect injuries, in a sense, predispose plants to root rot and also greatly complicate the development of varieties of cereals resistant to root rots.

Physiologic Specialisation is a common phenomenon among parasitic fungi, including those that cause root rots of cereals. Not only are the root rots of cereals caused by many distinct species of fungi but most of the species comprise many parasitic races. For example, Helminthosporium sativum consists of many distinct cultural and parasitic races. The races vary greatly in their parasitic capabilities on cereals and grasses. Some races of H. sativum are very virulent on wheat, barley, and rye. Others are moderately virulent. Still others are only weakly virulent. They also differ in host range. Some attack wheat, barley, rye, and oats. Others do not attack oats. Some attack corn readily. Others are nonvirulent on corn.

Similar genetic variation in parasitism has been reported in other root-rotting fungi Calonectria graminicola, Fusarium culmorum, F. graminearum, Ophiobolus graminis, and Rhizoctonia solani. Root-rotting fungi are not static. New races may arise by means of hybridisation between races or by Mutation. Differences in virulence of races are of great importance because they complicate the study of inheritance of resistance to root rots and the selection of varieties resistant to root rots.

New varieties may be resistant in one locality in North America, but susceptible in another because of the prevalence of different pathogens and the existence of different races of the pathogens in different regions. Moreover, root rot problems are continually changing because new pathogens are introduced or new virulent races of old pathogens are developed or introduced from the other regions or countries. Sometimes problems change because relatively minor root diseases become major diseases. Changes in cropping systems and the introduction of new varieties also may create new conditions. All new varieties therefore should be tested for resistance to root rot before they are made available to the growers.

One of the best means of selecting varieties resistant to root rots is to test them in pathological nurseries disease gardens in the regions in which they will be grown. Disease gardens were established in Minnesota in 1919 for testing the varietal reaction of wheat to diseases. The procedure is now common for testing different crops against root rot. A disease garden is a "sick" plot of land in which varieties of the same crop are grown in successive years and subjected to the most severe disease tests that can be devised. The soil in the disease gardens is inoculated with all available physiologic races of seed-borne and soil-inhabiting pathogens and potentially important root-rotting pathogens that occur in the region in which the varieties are to be grown. The gardens actually are the critical testing grounds for the selection of varieties and hybrids resistant to rot.

Different species of fungi or different physiologic races of some fungi may be widespread in some seasons but not others. If one wants to be certain of resistance, he must test the variety in the disease garden for several years. The production of new disease-resistant varieties takes at least 10 years, usually much longer. Root rot tests should be continued during that period.

The use of resistant varieties and good farming practices that promote strong plants are the best ways to reduce damage caused by root rots. Control measures generally do not eliminate the disease but tend to reduce the amount of infection. Specific control measures, therefore, are restricted at best to localities. To control or limit root rots one should plant sound seed; treat the seed with suitable fungicides; plant early, shallow, and at recommended rates; use good cultural methods; practice sanitation and good cropping sequence; and grow resistant varieties.

Seeds used to be treated with fungicides primarily to control smuts. Since 1949 many varieties of cereals have been developed that are resistant to smuts, so that treatment of their seed appears to be unnecessary. Nevertheless, seed treatment of the new smut-resistant varieties makes for increases in yield, chiefly because the newer fungicides are effective in preventing seed decay and the seedling blight.

Seedling blight can be controlled or reduced by treating seed with fungicides and growing the seed in relatively disease-free soil. The treatments retard primary seed-borne infection and thus reduce the amount of root rot inoculum early in the season. The early protection of the seedlings affords an easy way to increase crop yields. Seed treatments do not reduce the amount of secondary infection, as plants grown in infested soil may become infected any time between germination of seed and maturity.

Increases in yield of wheat, oats, and barley of 5 to 10 per cent or more are common. H. Murphy, of the Department of Agriculture, obtained increases up to 60 per cent in yields of oats when he treated seed heavily infected with H. victoriae with New Improved Ceresan and sowed it in experimental plots at Ames, Iowa. Deep-planted seeds of cereals are more subject to seedling blight than shallow-sown seeds.

That is especially true if seed lots are not disease-free, if the seedbed is not well prepared, and if weather conditions become unfavourable for normal germination. The deeper the seeds are planted, the longer the succulent tissues are exposed to attack by seed and soil-borne organisms.

Field and greenhouse tests indicate that deep seeding increases root rot and lowers stand and yield. Specific depth of seeding cannot be given, as the recommended depth of seeding varies with the crop, soil type, precipitation, and region in which the crop is grown. As a general rule, seed should be sown just deep enough to provide sufficient moisture for good germination.

Fertilisers apparently have little effect on the development of most root rots unless the fertility level is so low that the hosts do not grow well. Most attempts to control common rots by application of commercial fertilisers have been unsuccessful. On the other hand, H. R. Rosen and J. A. Elliott in Arkansas were able to control take-all by using commercial fertilisers.

S. D. Garrett in England recommended acid phosphates for the control of the disease. T. C. Vanterpool, in Alberta, Canada, found that an unbalanced phosphorus-nitrate relationship predisposes wheat seedlings to attack by Pythium.

Crop rotation and summer fallow often reduce damages from root rots, because many of the pathogens survive in the soil, on refuse of crops, and on wild and cultivated grasses. They also develop on plant remains in and on the surface of the soil. Some, such as species of Fusarium, attack crops other than cereals. Proper crop sequence is important therefore in fighting root rot.

Root rot is usually more prevalent in soils that have been cropped continuously to a given crop than those in which rotation has been practiced. A rotation that includes immune or highly resistant crops suppresses the pathogens and consequently reduces root rot. Take-all, for example, can be reduced materially if the crop sequence includes such nonsusceptible crops as alfalfa, sweetclover, flax, and corn. In regions where the race of the take-all fungus that attacks oats is not prevalent wheat can follow oats.

Summer fallow is also an effective control measure against take-all and to some extent against other types of root rots, with the exception of browning. In fact, summer fallow tends to increase the severity of this disease in wheat.

Antibiosis is the association of two kinds of organisms in which one inhibits the growth of the other. Antibiosis is important in the control of root rots. Agricultural soil is teeming with microscopic organisms algae, fungi, bacteria, protozoa, and nematodes. Some have no apparent effect on plant growth. Some are beneficial. Some, especially the root-rotting fungi, are destructive. As organisms grow and multiply rapidly in the soil and on plant refuse, they struggle constantly for survival. If it were not for this constant "biological warfare" the soil-borne organisms that cause root rots would multiply so fast that the arable land would become "sick" and then be unsuitable for grain crops.

Many factors other than the prevalence of micro-organisms in the soil determine how long a soil-borne pathogen can persist in the soil temperature, moisture, acidity, type of soil. Nevertheless, soil-borne pathogens cannot compete in the soil with their enemies, the saprophytic organisms.

Indeed, many pathogens that cannot survive in normal soil actually grow well in sterilised soil. Some of them can survive in the soil for a short time. Others may live in the soil for years. Those that persist are more tolerant to antibiotic substances and therefore are more difficult to control by cultural methods.

The biological control of soil-borne disease is promising, but we still do not know the extent to which the method can be applied. Adding large amounts of specific antibiotic organisms or their extracts to the soil will control or greatly reduce some soil-borne diseases. Seedling blight of wheat, barley, and rye caused by Helminthosporium species can be controlled by applying cultures of antibiotic organisms to the soil at the time of planting.

That practice may not be feasible on a large scale; cultural practices must then be resorted to. We have some evidence that antibiotic substances may be used as seed disinfectants for controlling soil-borne diseases. The application of manure and plant refuse to the soil and cropping sequences of cultivated crops have a pronounced influence on the number of micro-organisms in the soil. Take-all has been controlled by applying manure to the soil. More studies are needed on control of soil-borne diseases by means of antibiosis, especially in relation to crop sequence.

Varietal Resistance is perhaps the most promising key to the control of root rot diseases, even though resistance is only relative. The degree of susceptibility of a given variety is influenced tremendously by ecological conditions and cultural practices. Nevertheless, varieties differ greatly in susceptibility to specific organisms when grown under uniform tests.

Most varieties of wheat are susceptible to root rots. Apex, Thatcher, and Marquis are among the more resistant spring wheats to the common root rots. Kota, Kubanka, and McMurachy Selection are moderately resistant to take-all. All varieties of oats that have Victoria as one parent, such as Vicland, Tama, and Boone, are very susceptible to helminthosporium root rot. Many commercial varieties, such as Clinton, Benton, and Bonda, are highly resistant. A variety may be resistant to one type of rot but susceptible to another. Kubanka wheat is moderately resistant to take-all, but is susceptible to the common root rots.

Victoria oats is fairly resistant to root rot caused by Fusarium culmorum but extremely susceptible to H. victoriae. Peatland barley is moderately resistant to helminthosporium root rot but very susceptible to pythium root rot. Resistance must be bred into the adapted commercial varieties that do not have it.

That was done by scientists in Minnesota, who developed varieties of wheat, barley, and oats that are resistant to helminthosporium root rot. The disease gardens have been of practical help in eliminating inferior varieties and hybrid lines. The idea could well be expanded: Growing a world collection of wheat, oats, and barley in nurseries in different regions of the United States and maybe in other countries would give us a wonderful opportunity for selecting disease-resistant varieties and desirable parental materials.

In Conclusion: Root rots are among the most destructive diseases of cereals in the United States. They are debilitating, insidious, and usually inconspicuous, but sometimes they become epidemic. Their importance varies greatly with season, locality, and the crop. Host plants are subject to attack from the time the seeds are sown until the crops are mature.

The pathogens are of diverse types and attack all underground parts of the plants and induce root decay, seedling blight, and premature death of older plants. Root-rotting pathogens live in the soil and tend to increase in the soils when susceptible crops are grown year after year. Soil is not just soil; it is actually teeming with microscopic life protozoa, bacteria, and fungi.

Those organisms are carrying on a biological war, struggling for the survival of the fittest. They destroy many of the root-rotting organisms and hence help to prevent the development of root rot. Their relative prevalence in the soil can be greatly modified by cropping sequences and cultural methods.

Although root rots cannot be completely prevented where cereals are grown intensely, they can be greatly reduced by good agricultural practices. Sound seed of recommended varieties should be treated with a fungicide to eliminate the pathogens from the seed and to give the young seedlings protection against soil-borne organisms. Also, the seed should be sown only deep enough to provide adequate moisture for germination.,

A good cropping sequence must be followed. The seed beds should be well prepared, and appropriate fertiliser applied if there is a deficiency of essential elements. Varieties differ in susceptibility and only the recommended varieties should be grown. In general, good farming is the most feasible method of controlling root rots of cereals.

THE RUSTS OF WHEAT, OATS, BARLEY, RYE

Wheat, oats, barley, and rye may be attacked by eight distinct species or subspecies of rust fungi. Wheat is subject to stem rust (Puccinia graminis tritici), leaf rust (P. rubigovera), and stripe rust (P. glumarum). Oats are attacked by a stem rust (P. graminis avenae) and by crown rust (P. coronata avenae). Barley may fall prey to the same stem rust (P. graminis tritici) that attacks wheat and a leaf rust (P. hordei). A stem rust (P. graminis secalis) and a leaf rust (P. rubigovera) attack rye. Each of the eight rusts is made up of several or many -different races, which may attack certain varieties of a particular cereal crop but not others.

Stem rust of wheat causes the most spectacular and perhaps the greatest losses. Leaf rust of wheat and crown rust of oats occur more frequently, usually affect larger acreages, and so may cause greater average losses year in and year out. Before the extensive Use of resistant varieties, stem rust was most destructive in spring wheat in the northern Great Plains. Severe losses sometimes occur in the southern Plains, many Eastern States, California, and occasionally in localities in the Pacific Northwest and Intermountain States.

Leaf rust of wheat and crown rust of oats occur wherever wheat and oats are grown. They cause little damage west of the Rocky Mountains or in the drier parts of the Great Plains. Some damage occurs in the eastern half of the United States nearly every year. Reductions in acre yields are especially great in the Southern and Southeastern States although neither wheat nor oats is grown extensively in much of that area.

Most varieties of barley, some of oats, and all varieties of rye escape considerable damage because of their early maturity as compared with wheat. Stem rust sometimes injures barley seriously. Leaf rust is not usually serious but is sometimes locally destructive on spring barley. The stem and leaf rusts do not cause serious losses in rye. Stripe rust occurs most commonly on wheat but only in special situations in the United States has it caused serious losses.

Upper left: Section of wheat stem rust pustule with spores in the wintering stage. Upper right: Section of a pustule of wheat stem rust with red, summer-stage spores. Lower left: Wheat leaves infected with the summer stage of the leaf rust fungus. Lower right: Wheat stem infected with the red or summer stage of stem rust fungus.

Wheat stem rust is characterised by pustules that develop and break through the surface of the stems, leaves, and sheaths and often the chaff and beards of the wheat plant. Myriads of brick-red spores escape from the pustules and are carried by the wind to other wheat plants.

The crop is damaged by the growth of the rust fungus on the wheat stems and leaves and by the developing spores, both of which use up the water and nutrient materials needed for developing the wheat kernels. The water requirement in rusted wheat is much higher than in healthy wheat.

As a result, the kernels are badly shriveled, many of them being so light and chaffy that they are blown out with the chaff in threshing. The remaining grains may be shrunken to one-half or two-thirds normal sise. Losses range up to 85 or 90 per cent; at that point the crop is not worth harvesting and is a total loss. The rusted straw turns brown, becomes dry and brittle, and soon breaks over.

Wheat stem rust also attacks barley and occasionally rye. It attacks many wild grasses, including wild barley or squirreltail grasses (Hordeum species); certain wheatgrasses (Agropyron species); wild-rye grasses (Elymus species); bottlebrush grasses (Hystrix species); and some bromegrasses. It does not attack oats.

The rust lives over the summer on volunteer grains and wild grasses in the Southern States and in northern Mexico. These spores and those blown down from the north in late summer and early fall infect fall-sown wheat or barley. The rust lives over winter in the red rust stage in the southern part of the United States and in northern

Mexico but not in the Northern States. If weather conditions are favourable in the spring, the rust multiplies and the spores sweep northward with the advance of the crop season. Thus a heavy rust epidemic in Texas is a threat to the wheatfields of Oklahoma and Kansas, and the latter, in turn, are sources of inoculum for the grain-fields in the North Central and Intermountain States.

The red rust spores that spread the disease are about one-thousandth of an inch long. They fall upon a wheat leaf and may germinate in an hour in warm, humid weather or in several hours at temperatures of 40 to 50 F. Germinating spores send out germ tubes, which grow along the surface of the leaf or stem of the plant until they reach a breathing pore (stoma), where they enter, send out branches that grow within the tissues of the plant, draw nourishment from them for a week or longer, and then produce the red pustules with another crop of spores.

At low temperatures, or when there is but little sunlight, it may take 2 or 3 weeks from the time the germ tube first enters the plant until the spores form. In warm, moist weather, however, they may reach full development within a week. Thus a new generation of rust spores may be produced every 10 to 14 days during the spring and summer, starting in Texas and advancing northward with the progressive development of the wheat crop at different latitudes. Since a single rust pustule may produce 350,000spores, the rust can spread very rapidly.

Rising air currents can lift the spores up to altitudes exceeding 10,000 feet and from such heights winds can carry the spores many miles before they fall to the ground. A heavy rust epidemic on the nearly 4 million acres of wheat in South Dakota could produce about 2 sextillion rust spores. If only one in 10,000 of the spores blew north into North Dakota, four spores would be provided for every wheat plant in the State.

Spores carried northward reach young wheat plants that are in a succulent stage, in which they are easily infected with rust. Spores blown southward in early summer fall on ripening or ripe wheat, which is not readily infected with rust. Those blown southward in late summer can attack volunteer wheat, early-sown wheat, and certain grasses. These plants in turn serve as a source of inoculum for fall-sown wheat in the South, where the rust lives over winter.

An additional source of rust menaces the wheat in the northern half of the country rust that develops on the barberry. The production of the brick-red uredospores ceases as wheat approaches maturity and black spores (teliospores) are produced in the same pustules. The latter stage is important only in the Northern States because the spores cannot survive the hot summers in the South. The teliospores usually will not germinate immediately after they are formed but require a relatively long resting period, somewhat like the hard seeds of alfalfa and clover. They are not blown about by the wind, but remain on straw or stubble throughout the winter and germinate in the spring, especially in moist, cool weather.

On germination, they produce small colorless spores (sporidia), which germinate and readily infect certain species of barberry. Others are wild native species, Berberis canadensis in the eastern Allegheny region and B. fendleri in the Rocky Mountains. The Japanese barberry (B. thunbergii) and the ornamental evergreen barberries are not attacked.

The winter spores themselves do not directly cause infection on grains; their sporidia cannot infect grains or grasses, but infect only the barberry. The sporidia germinate on barberry leaves and send out small germ tubes that penetrate the epidermis directly. Only young parts of the plant therefore can be infected, for the germ tubes cannot penetrate the older, tougher parts.

Within a week or 10 days after germ tubes of the sporidia have entered the barberry, small yellowish or honey-colored spots appear on the infected parts, especially the young leaves. The yellow spots comprise the spermagonia, which contain the spermatic. Spermatic function in connection with the sexual reproduction of the rust by means of pycniospores, which may develop new races of the rust as a result of new combinations formed during sexual reproduction.

Within a short time the cluster cups (aecia) are produced, commonly on the under surface of the infected leaf. The small cup-shaped structures contain long chains of cluster-cup spores (aeciospores), which are shot forcibly from their cups, especially during moist weather, are blown about by the wind, and initiate the red, or summer, stage of the rust on grains or grasses.

The typical life history of stem rust where barberries become infected is, then, as follows: The rust overwinters in the black, or teliospore, stage; the teliospores germinate in the spring and produce sporidia, which cannot infect grains and grasses but do infect certain kinds of barberries, on which the aerial stage subsequently develops. The aeciospores cannot infect the barberry but, falling upon grains and grasses, germinate and induce the uredial stage.

This stage may persist and produce successive crops of uredospores until growing conditions become unfavourable, when the teliospores are produced. While the stem rust fungus actually produces five kinds of spores, only the uredospores and aeciospores can infect grains and grasses. The teliospores, by means of which the rust usually survives the winter, are harmless unless there are barberries nearby.

Usually the rust is spreading from barberries to the grainfields some 2 or 3 weeks before the general rust spread arrives from the South. Because of this early start they may cause severe damage to grains growing near infected barberry bushes. The spread from barberries usually can be traced only short distances or up to a few miles; so the damage is largely local. However, they supply additional inoculum to hasten the spread of general epidemics. As many as 70 billion aeciospores may be produced on one large barberry bush.

About 240 parasitic strains or physiologic races of stem rust of wheat have been discovered, but not more than a dozen are widely prevalent or are important in any one year. Even fewer recognised races are of common occurrence in South America. These races differ in their ability to attack certain varieties of wheat. For example, some attack Marquis but not Pawnee; others attack Pawnee but not Marquis.

The Little Club variety is attacked by all races. Race 15B attacks all wheat varieties grown on farms in the United States, but does not attack certain varieties from Kenya, Africa. The same parasitic strains are not always present in a given region in different years, and there may be different strains in several regions in the same year. For that reason a variety may be susceptible to rust in one region and resistant in another; it may be susceptible in a given region in certain years and resistant in others.

The development of stem rust on wheat, then, depends not only on the presence of numerous spores early in the growing season but also on the presence of spores of the particular parasitic strains that normally can attack the variety of wheat grown in that region. Conversely, the particular races that occur in any region depend in part on the variety or varieties of wheat that are grown.

The Leaf Rust (Puccinia rubigo-vera tritici) of wheat usually attacks the leaf blades and sheaths, although it may sometimes occur on the stem proper, especially just below the heads. Sometimes it occurs also on the chaff' and awns or beards. The rust pustules are smaller than those of stem rust, are more nearly round, and are less likely to unite. Usually they appear only on the upper side of the leaf. The colour of the summer (uredial) stage is orange to orange brown, being brighter than that of pustules of stem rust.

The pustules of the black, or winter, stage (telia) are of about the same sise as those of the red stage, but they seldom break through the epidermis of the plant, and the colour, therefore, is likely to be lead gray. Leaf rust of wheat is found in nearly every place that wheat is grown. In the United States it is most abundant and destructive in the Southeastern States and in the Ohio and Mississippi Valleys, where weather conditions usually are most favourable for its development. It is less destructive in the hard red winter and the hard red spring wheat areas, although it is usually present to some extent and in favourable seasons may cause considerable damage to certain varieties. It usually is present in the irrigated areas of the West and on the Pacific coast but seldom does extensive damage, although sometimes it may be of importance locally.

Leaf rust may cause much damage to certain varieties of wheat in certain regions, although the damage usually is less conspicuous than that caused by stem rust. Leaf rust seldom shrivels the kernels, but it does reduce their sise and number and the quality of the grain. Furthermore, when plants become heavily rusted while still young, the entire plant may be weakened and somewhat dwarfed, and, under such circumstances, the yield may be reduced as much as 90 per cent.

This rust may occasionally attack barley to a slight extent, but for practical purposes it may be considered as attacking only wheat and a few species of goatgrasses (Aegilops species). The cluster-cup, or aecial, stage of leaf rust occurs on certain species of meadowrue (Thalictrum), but this is of no practical importance in the United States, as the species that are native here do not get infected under natural conditions.

The many parasitic strains of the leaf rust of wheat differ in their ability to attack different varieties. More than 140 of these strains are known and 85 have been identified in the United States. New ones are occasionally encountered. The exact manner in which they originate, however, is not known.

Leaf rust of wheat, like all other rusts, is caused by a very small parasitic fungus that enters the wheat plant through its breathing pores. The red, or summer, stage continues as long as the wheat plants are green and growing. The orange-colored pustules that develop on the wheat contain thousands of summer spores (uredospores), which are easily blown about by the wind and are capable of causing rust on other wheat plants.

During warm, moist weather, this red stage may recur every week; therefore the rust can increase and spread rapidly. When conditions become unfavourable for the growth of the wheat and for the development of the red stage, the black, or telial, stage appears. The red stage is capable of surviving the winter in most of the wheat-growing regions of the United States. It can be found at all times of the year in the Southern States and in some years in the Northern States.

If weather conditions in late summer and early fall favour rust development that is, if there is considerable moisture the red stage attacks volunteer wheat plants and fall-sown wheat, where it may remain and even increase somewhat during the winter.

In the spring it increases rapidly if weather conditions are favourable. The spores then may be carried long distances by the wind and infect wheat plants in regions where the rust may not have survived the winter. The destructiveness of the epidemic will depend on the earliness with which plants are attacked in the spring.

When there is abundant rainfall or heavy dews in the spring and early summer, following a winter that has favoured the survival of the red stage of the rust, abundant early infection may occur. If the weather continues warm and moist, there is danger of a destructive epidemic. As in the case of stem rust, warm, moist weather is most favourable for the development of leaf rust, but this rust seems able to develop under a wider range of conditions than does stem rust.

Stripe rust, Puccinia glumarum, has been commonly known as yellow rust because of the yellow or orange-yellow pustules of the summer stage. Stripe rust seems a better name, however, because one of its main features is the arrangement of the pustules in rows of various lengths, giving the appearance of fairly narrow yellow stripes.

Stripe rust attacks wheat, barley, rye, spelt, emmer, and more than 60 species of wild and cultivated grasses. In the United States it is most prevalent on wheat and certain wild grasses. It is seldom found on barley and rarely on rye. It is especially common on some of the wild barleys and bromegrasses and on a species of goatgrass.

Stripe rust develops most abundantly on the glumes or chaff, on the leaves, and on the leaf sheaths, but it may also attack the stems and the kernels. On seedling plants, and sometimes on older plants, the stripes are not distinct, but the yellow colour distinguishes the rust from other cereal rusts.

If the attack of stripe rust is heavy, especially on the necks and glumes, considerable damage is likely to result, particularly if plants have become rusted during the milk stage or earlier. Under such conditions the kernels may be shriveled and the yields considerably reduced. Often, however, stripe rust is most abundant in seasons that are so favourable to wheat that high yields are obtained despite the damage from the disease.

The black stage of the rust is formed after the red, or summer, stage but may develop at any time during the growing season and at any stage in the development of the host. The black pustules (telia) also are produced more or less in rows that look like long, narrow, dark-brown, or black stripes. They may appear on all above-ground parts of the plant, including even the kernels. Stripe rust is the commonest and most destructive grain rust in many regions of Europe. It occurs also in Africa, South America, Japan, China, and India.

Stripe rust has been in the United States at least since 1892 but was not recognised until 1915. It occurs sporadically over the western half of the United States and at corresponding longitudes in Canada and Mexico. The several physiologic races of stripe rust differ in their ability to attack certain varieties of wheat and other grass species. A special race attacks only a wild barley (Hordeum murinum). Both the summer, or yellow-spore, stage and the black stage of stripe rust are produced in the United States.

An alternate host and therefore the cluster-cup (aecial) stage are unknown. The exact conditions under which the yellow stage overwinters most abundantly are not known, although it persists throughout the winter under a variety of climatic conditions. While stripe rust is extremely destructive in some foreign countries, it has caused relatively little damage in the United States, except in limited areas where susceptible varieties are grown and where weather conditions are very favourable.

Stripe rust has not become established in the principal grain-growing regions of the Mississippi Valley. Experience in other countries, however, indicates that it might be destructive to some varieties of wheat under favourable weather conditions. When the upper part of the stems, chaff, and kernels become rusted during the milk or early dough stage of kernel development, the grain may be considerably shriveled. Yields then are smaller and the wheat is of inferior quality. Badly rusted grain may germinate poorly, but infection is not transmitted to plants grown from such diseased seed.

Stem Rust On Oats (Puccinia graminis avenae) looks like the stem rust on wheat. Stem rust is darker in colour than crown rust and usually produces longer pustules and is more abundant on the necks of the plants. The colour of the summer stage of stem rust usually is brick red; that of crown rust is a bright yellow or yellow orange. This disease attacks oats, a number of the wild grasses, orchardgrass, some of the fescues, meadow foxtail, and bluejoint. Normally it does not attack wheat, barley, or rye. Its life history and methods of spread are like those of the stem rust of wheat.

Crown Rust Of Oats (Puccinia coronata), often known also as leaf rust or orange leaf rust of oats, occurs principally on the leaves, although it is frequently present also on the leaf sheaths and sometimes on the stems and panicles. The most destructive disease of oats, it often cuts the yields 20 to 50 per cent. Crown rust is particularly destructive in the South and in the North Central States. More than 80 species of grasses may be attacked by one or another of several varieties of crown rust. Among them are quackgrass, reedgrass, redtop, meadow fescue, ryegrass, and bluegrass. The variety of crown rust most prevalent on oats also attacks several grasses and a number of different kinds of buckthorn. The oats variety of crown rust comprises in turn a number of distinct parasitic strains, which differ in their ability to attack different varieties of oats. More than 100 physiologic races of crown rust have been identified in the United States.

The pustules of the red, or summer, stage of crown rust are usually more or less circular. They rupture the epidermis and release uredospores. Later in the season the black, or winter, stage appears. The black pustules usually do not rupture the epidermis. The cluster-cup stage of crown rust develops on a number of species of buckthorn (Rhamnus). The first stage (spermagonium stage) appears on the upper surfaces of the leaves or on young twigs as small bright-yellow or orange spots. Opposite these spots, usually on the under surface of the leaf, the cluster-cup stage appears.

The life history of crown rust is similar to that of stem rust except that the cluster-cup stage develops on buckthorns instead of on barberries. The summer spores overwinter in the South and are spread to other plants and fields in the spring. In the Northern States the red, or summer, stage does not overwinter commonly, and epidemics result from summer spores that may be blown in from the South or from the development of the cluster-cup stage on buckthorns.

Warm, moist weather is most favourable for the rapid development and spread of crown rust. Moisture from dews or rains is likely to be held longer in dense stands of oats; that favours the development of rust. Epidemics are most likely to develop when there are many spores in the spring and when weather conditions favour the development of the rust. In the fields of winter oats in the South, where the summer spores survive the winter, the rust gets an early start in the spring.

The amount that overwinters depends on the amount of infection in the previous season and on weather conditions; mild winters are particularly favourable for the survival of the rust. In the Northern States the amount of infection on buckthorn bushes determines how early and how good a start the rust gets.

If the weather in the early spring has been favourable for abundant infection on the buckthorns, there may be considerable crown rust if the subsequent weather conditions are also favourable. Summer spores blown into the Northern States from the South may also cause widespread epidemics. All buckthorn bushes in the northern oat-growing States should be destroyed because inoculum from them causes severe losses to oats in the vicinity and provides inoculum for an early spread of crown rust.

Barley is attacked by both the wheat and rye strains of stem rust and the development and spread of the disease on barley is the same as for those crops.

Leaf Rust Of Barley is seldom of economic importance in areas where spring barley is grown but is sometimes destructive locally to winter barley. At times it is so abundant as to prevent the proper heading of infected plants and may sometimes reduce the yields and the quality of the grain.

The pustules of the summer stage of leaf rust of barley appear on the leaf blades and sheaths of the barley plant. They are yellow or yellowish brown, small, and round. The black stage follows the red stage and produces lead-gray pustules, which do not rupture the epidermis of the plant.

Leaf rust attacks cultivated varieties of barley, although under some conditions it may develop weakly on some of the other cereals and on several wild grasses.

The life history of leaf rust of barley is similar to that of the leaf rust of wheat. The rust survives the winters in the red, or summer, stage, particularly in the winter regions. The spring, or cluster-cup, stage, when it is produced, develops on the star-of-Bethlehem (Ornithogalum umbellatum), and a closely related species (O. narbonense). Although those plants are fairly common in certain parts of the United States, they rarely become rusted.

Stem Rust Of Rye (Puccinia graminis secalis) has a life history similar to that of wheat rust. However, stem rust of rye attacks barley but does not develop on the wheat and oats, except under specially favourable experimental conditions. In addition, it attacks many of the same grasses as does wheat stem rust from wheat but particularly quackgrass, which is not commonly attacked by wheat stem rust. Because of the early maturity of rye and the fact that it is grown but little in the South, severe attacks of stem rust on rye are rare except where there are barberry bushes. The eradication of barberry is the only control measure needed.

Leaf Rust Of Rye (Puccinia rubigovera) is similar to the leaf rust of wheat; so similar, in fact, that it is considered as belonging to the same species. It may occur wherever rye is grown. In the South, where rye is sometimes used for winter pasturage, it may become so abundant as to kill the plants during the winter. It may also cause considerable losses in the northern rye-growing regions. It attacks Secale montanum, a wild species of rye. Some wild grasses may occasionally become infected if the conditions are favourable, but the rust generally is restricted to rye.

It usually attacks the leaf blades and sheaths. It may occur also on the necks and glumes of severely rusted plants. The pustules of the red stage are scattered more or less irregularly and are orange brown to cinnamon brown. They are usually small but may unite to form fairly large pustules. The pustules of the black stage are gray to black and remain covered by the epidermis until it decays or falls away. The rust persists during the winter in the red, or summer (uredial), stage. The spring, or cluster-cup, stage can develop on some plants of the borage family, but it is seldom found in nature in the United States and is therefore of no importance in the life history of the rust.

The development of epidemics occurs under conditions similar to those that promote epidemics of leaf rust of wheat. The degree to which the summer stage persists during the winter depends on weather conditions. In the spring, new infections occur; if the weather is warm and humid, the rust may become epidemic.

Rainy weather or cool, dewy nights and warm, humid days are most favourable for the development of the rust. No variety of rye is uniformly highly resistant to leaf rust. Rye is cross-pollinated and no varieties are uniformly pure. Certain pure inbred self-fertile strains are highly resistant to leaf rust but are not vigorous and are therefore useful only as resistant breeding stocks.

Three Methods Of Controlling leaf and stem rusts of wheat, rye, oats, and barley have been suggested the use of resistant varieties, eradication of the alternate host, and dusting with fungicides. Early varieties and such cultural practices as early seeding and use of phosphate fertilisers that hasten ripening may help to escape rust but do not prevent damage in bad years. Their effect is indirect rather than direct. In general the recommended control measures are much the same for both leaf and stem rust and for all small grains with some modifications, depending on the particular crop or alternate host.

Late seeding, especially of spring grains, delays ripening and is likely to increase the damage because it provides a longer period for the development of the rust organism. Farmers generally recognise the advantages of early seeding and usually do their seeding as early as possible. Phosphorus often hastens ripening and thereby reduces rust damage. Heavy applications of barnyard manure or of nitrogen, on the other hand, may delay ripening and also may produce a heavy vegetative growth that retains moisture and favours the development of rust. Extensive trials have shown that stem and leaf rusts of wheat and presumably those of other cereals can be prevented by dusting with sulfur or other suitable fungicides. Usually three or more treatments during the season are necessary.

More are needed if the fungicide is washed off by frequent rains. Dusting should begin before rust damage appears, but that is wasted effort in seasons in which rust development is light and the cost generally is too great to be practical in the United States. Some of the newer fungicides may be more effective than sulfur. Eradication of the common barberry (the alternate host of stem rust of wheat, rye, oats, and barley) and of buckthorn (the alternate host of crown rust of oats) has been widely practiced.

Damage from stem rust has been greatly reduced since eradication of the barberry was started. At the same time an aggressive breeding programme has sought to produce resistant varieties, especially of wheat. In the more humid sections of the northern Great Plains where stem rust has been most destructive, practically no varieties of wheat susceptible to the ordinary races of stem rust (other than race 15B) have been grown since about 1938. It is certain that eradication of the barberries did not prevent the widespread distribution of race 15B in 1950 and again in 1952 nor infection of susceptible varieties in experimental plots at several locations in recent years.

The discovery that rust may hybridise on barberry bushes and thus produce new races is an additional reason for continuing the eradication campaign. Varieties formerly considered to be resistant, including Ceres, Vernal emmer, and Iumillo durum, were attacked later by races of rust believed to have originated on the barberry. Despite this menace from new races, Iumillo durum retained its resistance to rust in the field in the United States for nearly 50 years and Vernal emmer probably for 75 years, and both for an unknown period elsewhere before finally succumbing to race 15B. We have evidence also that new rust races may be produced by mutations in the absence of barberry bushes. Whatever the answer may be, the past losses and future potential losses are so great that no promising method of control should be neglected. Eradicating the common barberry and breeding resistant varieties both fall in this category.

Breeding resistant varieties is certainly one of the most promising methods for controlling all rusts of small grains. It takes a long time, however, and often is difficult. Since stem rust has been most destructive on wheat, most of the outstanding examples of successful breeding for resistance to this disease relate to wheat. Outstanding progress also was made in breeding oats resistant to crown rust.

Both stem and leaf rust of wheat were recognised as among the more important hazards in growing wheat in many sections of the United States before 1900. About 1890 B. T. Galloway of the Department of Agriculture conducted spraying experiments that showed that rust could be reduced by fungicides, but he concluded that resistant varieties, if available, would be a more practical method of control. No markedly resistant varieties were known at that time.

M. A. Carleton, who was employed by the Department of Agriculture in 1894, sowed an extensive collection of wheats from many countries at Garrett Park, Md., that fall, in a search for resistant varieties. It was known then that soft winter varieties grown extensively in Texas, Kansas, and southern Nebraska were more susceptible than the hard winter wheats then being adopted in those States, but even the latter were severely damaged in years when rust was bad. Some of the durums were known to be resistant to leaf rust, but they had not been used to any material degree as parents in crosses.

By 1900 Mr. Carleton and others had introduced from Russia and other countries a number of varieties of durum wheat and emmer, which were widely grown in experimental plots throughout the Great Plains. In 1902 and again in 1904 at Brookings, S. Dak., John S. Cole observed that certain varieties, specifically Yaroslav emmer from Russia and lumillo durum from Italy, were almost completely immune to attacks by stem rust.

They were not satisfactory commercial or bread-making wheats, but they provided rust-resistant parents for hybrids and thereby the basis for an effective breeding programme. Their usefulness is attested by the fact that practically all varieties grown in 1953 in Minnesota, the Dakotas, Wisconsin, and Canada, many of those grown in California, and some of those grown in the Eastern States carry resistance genes from one or both of those two early introductions. The job has by no means been completed as proved by the widespread invasion of race 15B, which attacks all commercial varieties.

Losses from rust have been greatly reduced; even more important perhaps are the lessons that have been learned and the confidence in ultimate success that has been generated as a result of this experience. Such progress as has been made is the result of more than 60 years of research. Many difficulties have been met and overcome. A brief account of some of the problems and the ways in which they have been solved may be of value.

Crosses between commercial varieties of common wheat and Iumillo durum and Yaroslav emmer were made by Mr. Cole at Brookings almost immediately after the resistance of the latter was discovered in 1904 or 1905. But the crosses were lost before anything useful was derived from them because of changes in personnel. Similar crosses were made at the University of Minnesota about 1907 and at other places in later years.

Serious difficulties were at once encountered because of a high degree of sterility and of linkage of rust resistance with undesirable characteristics from the durum and emmer parents. As a result it was nearly 30 years before any really satisfactory variety was derived from such crosses. Other sources of resistance had been discovered in the meantime and used in crosses to produce some commercially desirable common wheats.

The first of the latter to become really important was Ceres, produced by L. R. Waldron of the North Dakota Agricultural Experiment Station and distributed to farmers in 1926. Ceres was derived from a cross between the famous Marquis variety from Canada and Kota, a rust-resistant common wheat found mixed in a durum wheat that had been introduced from Russia by H. L. Bolley in 1903. Ceres was resistant to the prevailing races of stem rust. It was also moderately early and resistant to drought. It was soon grown throughout the northern Great Plains until after 1935, the year in which it was severely damaged by race 56, to which it is susceptible and which first appeared in epidemic proportions in that year.

The second resistant common wheat to be developed from a cross was Hope. It was first distributed to farmers in 1927. It was never widely grown, largely because of its susceptibility to heat and drought and low relative yields in the absence of rust. It is of great interest, however, because it or its near relatives have entered into the parentage of most of the resistant varieties now generally grown (except Thatcher) and because of the somewhat unorthodox manner in which it was produced.

Hope was derived from a cross between Yaroslav emmer and Marquis wheat made in 1915 by E. S. McFadden, then employed by the South Dakota Agricultural Experiment Station as an undergraduate student assistant. Because of the high degree of sterility, only a few poorly developed seeds were secured from the cross. From them developed a single plant.

The first of the subsequent generations consisted of nondescript, unpromising plants, which were grown in bulk at the Highmore Substation in order "to permit natural selection to eliminate some of the undesirable combinations. . . ." An important feature was the strict elimination of shrunken seed by screens and strong blasts of air in a fanning mill. Another was the relatively large population of plants which could be grown by this method. Marquillo was the first commercially grown variety to be derived from crosses involving Iumillo. It was distributed by the Minnesota Agricultural Experiment Station in 1928. Like Hope, it was never extensively grown, mostly because of the yellow colour of the flour produced from it. It and sister strains have been useful, however, as parents in other crosses.

The famous Thatcher variety, distributed by the Minnesota Station in 1934, was derived from a double cross involving a sister strain of Marquillo and a selection from a cross between Marquis and Kanred winter wheats. As Kanred is resistant to some races of stem rust, it is generally considered that the resistance of Thatcher is due to genes both from the Iumillo durum and from Kanred.

Thatcher was first grown by farmers in 1934. In the severe stem rust epidemic of 1935 it often outyielded Marquis and Ceres by 20 to 30 bushels an acre. In subsequent years its acreage increased as rapidly as seed supplies permitted. It was highly resistant to the prevailing races of stem rust, ripened early, had short, stiff straw, and produced grain of excellent quality. By 1939 it was the most important variety of spring wheat in the United States.

Thatcher, however, is very susceptible to leaf rust. It was severely damaged by that disease in 1938 and 1941-In the meantime, other new varieties resistant both to leaf and stem rust had been produced and distributed to farmers. The principal ones are Rival and Pilot, distributed in 1939; Regent, released in Canada in 1939; New-thatch, released in 1944; Mida, in 1944; Rushmore, in 1949; and Lee, in 1951. The acreage of Thatcher began to decline after 1940 and since then has practically disappeared in the eastern half of the northern Plains of the United States, although it continued to be grown in the drier western sections and in Canada.

All of the newer varieties, except Lee, derived most of their resistance from Hope or H-44, a sister selection of Hope. Hope and its derivatives fortunately were resistant to the prevailing races of leaf rust as well as stem rust. About 1944 a new race of leaf rust that attacked most previously resistant varieties appeared generally in farmers' fields throughout the Great Plains. That meant that varieties resistant to the new race and to the older races were needed.

Lee was produced in Minnesota from a cross between Hope and Timstein. Timstein is a selection from a cross between Triticum timopheevi, a wheat relative, and Steinwedel, a common wheat, made by J. T. Pridham of Australiaand brought to the United States by his associate, S. L. McIndoe. Lee is highly resistant to most prevailing races of leaf and stem rust except stem rust race 15B. Interest in 1953 centreed in the production of new varieties resistant to 15B as well as to other races of stem rust. The race now known as 15B was identified as early as 1939, and attention was at once directed to discovering new sources of resistance. Since the widespread onset of 15B in farmers' fields in 1950, these efforts have been greatly intensified.

Much progress has been made and it is reasonably certain that new varieties will be produced that are resistant to all the prevalent races both of leaf and stem rust and which are satisfactory or superior in other respects. Considerable control of stem and leaf rust in the southern Great Plains has come about indirectly, and to some extent perhaps unexpectedly, through the development of early maturing varieties. Many of the newer ones carry genes for resistance derived from various sources. Even more important has been the discovery that early ripening varieties are generally superior throughout much of the region even in the absence of rust. Unfortunately these early varieties are not so winter hardy as the old Turkey and others they largely replaced and hence are not suitable for the colder areas.

Some notion of the importance of this development may be gained from the fact that early or moderately early varieties, including such important ones as Pawnee, Comanche, and Wichita, occupied more than 75 per cent of the wheat acreage in Kansas in 1952. Breeding for resistance to leaf rust and in some cases also for resistance to stem rust has been an important objective in several Eastern States Indiana, Illinois, Kentucky, Georgia, and North Carolina and by the Department of Agriculture at Beltsville, Md. Several new varieties resistant to leaf rust and in some cases to stem rust have been released to farmers.

The more important of these are Vigo in Indiana; Saline in Illinois; Chancellor in Georgia; Atlas 50 and Atlas 66 in North Carolina; Anderson in South Carolina and North Carolina; and Coastal in South Carolina. The history of breeding for rust resistance in the durum wheats is similar to that for the hard spring, but is less involved. Most of the introduced durums at the time of their release were moderately or highly resistant to the prevailing races of both rusts.

New or previously unimportant races, especially stem rust races 17 and 21, to which they are susceptible, soon appeared, however. Among the earliest attempts to obtain resistant varieties were crosses made in Minnesota between durums of good quality and the poor but highly resistant red durum, Pentad. No suitable progenies came from the cross, and attention was then concentrated on Vernal emmer as a source of resistance. Stewart and Carleton, produced in North Dakota from a Mindum × Vernal emmer cross and distributed in 1943, are resistant to these and older races. Stewart has since become the leading variety of durum wheat. Vernum and Nugget, also produced in North Dakota, were released to farmers in 1947 and 1951, respectively. None of these newer varieties nor the old ones is resistant to 15B. Resistance to 15B has been found in certain poor-quality durums from Portugal and Spain, and they have been crossed with the leading commercial varieties. Promising selections are being increased.

All commercially grown varieties of durum are even more susceptible to 15B than are the common wheats. Moreover, the durums generally ripen later and hence favour the development of this race of stem rust to an unusual degree. Although the immediate outlook is rather unfavourable, we believe that satisfactory resistant varieties will be developed.

The experience since 1900 helps to explain why the creation of new varieties is often a slow and uncertain process. Sterility and linkage are always a problem in dealing with intergeneric crosses, but plant breeders know better how to deal with them than they did a generation ago. In Oats more than in any other major crop, there has been a parade of new varieties and frequent and complete changes in varieties in the past 50 years. Breeders have been successful in producing high-yielding, rust-resistant varieties widely grown by farmers. As new or previously unimportant races of the rust organisms have appeared, still other varieties resistant to them have been developed. The introduction and creation of early maturing varieties that escape much rust damage have also been important factors.

At the beginning of the century only late or medium-early varieties and only rust-susceptible varieties (with a few exceptions) were available. The main exceptions were the late or comparatively late White Russian and Green Russian varieties, which were resistant to stem rust but highly susceptible to leaf rust, and Red Rustproof and Burt, in the Southern States. It is now known that the Red Rustproof types are better designated as late rusters rather than as rustproof, since they are not resistant.

The first important step in avoiding damage from oat rusts in general consisted of the introduction of the Kherson and Sixty Day varieties from Russia, the first by the Nebraska Agricultural Experiment Station in 1896 and the second by the Department of Agriculture in 1901. Neither is resistant to stem or crown rust, but they matured early and escaped much damage that otherwise would have occurred. Widely adapted throughout the central United States from the Atlantic almost to the Pacific coast, they soon became the dominant varieties, especially in the Corn Belt. Another important introduction was Swedish Select, brought to the United States from Russia in 1899. It also is not resistant to either stem or crown rust but matured somewhat earlier than many of the varieties it replaced in the northern United States. Burt, an early-maturing crown-rust-escaping selection from Red Rustproof, made in 1878, has been widely grown throughout the South and sparingly in the southern part of the Corn Belt.

Another important early but rust-susceptible variety was Fulghum, selected about 1900 from Red Rustproof by a farmer, J. A. Fulghum of Warrenton, Ga. Kanota, often regarded as a synonym of Fulghum, was distributed in 1919 by the Kansas Agricultural Experiment Station. Because of its early maturity, it did much to reduce crown rust damage in Kansas and nearby States.

The importance of breeding for resistance to stem rust apparently was first recognised about 1918 when White Russian was crossed with Victory in Minnesota. From it the resistant variety Anthony was produced. Anthony was distributed to farmers in 1929. Richland and Iogold were Selected at the Iowa Agricultural Experiment Station from Kherson in 1906 and distributed to farmers in 1914 and 1926, respectively. Rainbow, highly resistant to stem rust and moderately resistant to crown rust, was selected at the North Dakota Agricultural Experiment Station in 1925 and distributed to farmers in 1930.

One of the first definite attempts to produce varieties resistant to crown rust was made in 1928 when Rainbow, resistant to stem rust and moderately resistant to crown rust, was crossed with Markton at Aberdeen, Idaho. Selections later were grown at Ames, Iowa, and Arlington Farm, Va., where they were tested for resistance to rust, for yield, and for other characteristics. One of them, Marion, was distributed to farmers in 1941.

Several severe epidemics of crown rust and the discovery of the high degree of resistance of two varieties to the disease centreed the attention of oat breeders on crown rust resistance as an important objective. One, Victoria, had been introduced from Uruguay in 1927. The other, Bond, came from New South Wales in 1929. Neither was satisfactory for growing on United States farms but appeared to be just what the oat breeders wanted as parent material.

Victoria was crossed with Richland in 1930. In subsequent years 30 varieties were selected from it and other Victoria crosses and distributed to farmers. They were soon the dominant varieties in all Corn Belt States and were important also in the Northeast and South. They were grown on approximately 30 million acres about two-thirds of our oats acreage in 1946. They produced high average yields, had short, stiff straw, and did not easily lodge, a characteristic almost unique among varieties adapted to the Corn Belt and highly prised by farmers who use combines. Helminthosporium blight, a previously unimportant disease, caught up with the oats crop in 1946. Victoria and selections from crosses in which Victoria was a parent were especially susceptible and were seriously damaged. Fortunately Bond was resistant; in the meantime it had been used extensively as a parent and a number of selections were then in the yield-testing stage. Seed supplies of a few of them had been increased to be distributed if and when their superiority should be demonstrated.

The widespread damage from helminthosporium blight emphasised their potential value and stimulated more extensive tests for yield and other characteristics. Out of this work have come such outstanding varieties as Clinton, Bonda, Andrew, Benton, and Mindo, now widely grown in the Corn Belt; and Camellia, Taggart, and Delair, which are grown to some extent in the Southern States. These new varieties, derived from crosses in which Bond was one parent, are resistant to the helminthosporium blight and generally to crown rust and are fully equal to the Victoria-derived varieties in yield, stiffness of straw, and quality.

Those achievements do not, however, justify complacency. Crown rust races 45 and 57 attack Bond and most of its derivatives. More recently crown rust race rot has been discovered; it attacks both Victoria and Bond and their derivatives. Stem rust came again into the picture with races 6, 7, and 8, to which most of the new varieties are susceptible. Race 8 is widespread and abundant. Race 7 appears to be rapidly increasing. A new, more virulent biotype, 7A, has been reported from Canada. Race 6 occurs infrequently. Segregates from crosses designed to combine resistance to those races with resistance to other races of stem rust, crown rust, and the helminthosporium blight were in the yield-testing stage in 1953.

Rusts of barley were not generally considered important enough to justify intensive breeding for resistance to them until about 1935, when many fields in the North Central States and western prairie regions were severely damaged by stem rust. It is now generally recognised that leaf rust may do damage generally in all humid barley-growing areas but principally on winter barley in the Eastern and Southeastern States. Resistance to both rusts has been an important objective in most barley-breeding programmes in those areas.

The most important commercial variety resistant to stem rust was found by a farmer, S. T. Lykkens of Kindred, N. Dak. In 1935 he decided to plow up his field of Wisconsin Pedigree 37 because of severe damage by stem rust. He observed a single plant that was free of rust, saved the seed, planted it in his garden the next spring, and in subsequent years sold seed to his neighbours. By 1942 it was a recognised commercial variety. Its resistance has since been verified by others. It has been included in numerous yield trials and tests of quality. Although it has some defects, such as weak straw, it has compared favourably with other varieties in most important respects. It has been named Kindred and is now grown more extensively for use in malting than any other variety in the principal barley-producing area of the United States.

Other commercially grown varieties resistant to stem rust are Peatland, Mars, Plains, Feebar, and Moore. Fifty or more other varieties, most of them from the world collection of barley from various foreign countries and as selections from farmers' fields known to be resistant to stem rust, are valuable sources of resistance for use by plant breeders, even though they are not grown commercially. Progress has been made also in breeding varieties resistant to leaf rust, but only one, Goliad, had been released in 1953. Goliad is also resistant to stem rust.

The large number of physiologic races (240 or more of stem rust of wheat, for example) has sometimes seemed an insurmountable obstacle to the successful breeding of resistant varieties. Nevertheless a great deal of progress has been made. From the years of experience has come a better understanding of their relation to breeding programmes. Our concept of a large number of races stems mostly from the manner in which they are identified. Because of technical difficulties, it is not generally feasible to determine the ability of each of a large number of collections of rust spores to infect each of a large number of varieties in an advanced stage of growth in the field. Yet this is the information that is needed.

The nearest approach to this so far devised for identifying races is to infect young plants, usually seedlings, growing in a greenhouse. Only a few varieties, often not more than a dozen, are included in the tests, and they are usually the same year after year. They are known as host testers. It is hardly to be expected that results from the tests could be applied directly to breeding without due consideration of the conditions under which they are obtained.

One important fact is that resistance or susceptibility in the seedling stage in the greenhouse does not necessarily mean resistance or susceptibility when a plant is in an advanced stage of development in the field. A physiologic race that infects a given variety in the seedling stage but not when the plants are approaching maturity is not likely to be a serious menace to that variety, but it may add to the number of physiologic races. Another fact that makes breeding for resistance to all prevalent races less formidable than it may seem to be is that many varieties are resistant to several races. It is known, for example, that Kanred in the seedling stage is resistant to 11 races of stem rust. Hope is resistant to at least 17 races, Red Egyptian to 26, Kenya-Gular to 30, and certain other Kenya derivatives to at least 35. They may be and probably are resistant to a larger number in advanced stages of growth in the field, although, as we indicated, the exact situation would be hard to determine.

Somewhat the same applies to leaf rust of wheat and to the rusts of oats and barley. In most cases it is possible on the basis of known information to choose a small number of varieties of each crop that collectively are resistant to all known races of a given rust. If this resistance were to be concentrated in a single variety, and no new races appeared, the job of producing resistant varieties would be finished except as there might be need for resistant varieties for other areas or for other reasons. And this, of course, is what cereal breeders for rust resistance have been trying to do for some 75 years with considerable success.

LEAF AND HEAD BLIGHTS OF CEREALS

Leaf blights reduce the green leaf surface that a plant needs to produce its carbohydrates. If that happens early in the life of a cereal plant, it has fewer numbers of heads and kernels in a head. Later leaf blighting reduces the sise and weight of kernels.

Head blights kill the head or many of the young kernels when the attack occurs early. When the attack occurs after the kernels are partly developed, the grain is shriveled and may be lost in threshing. Blight late in the season may cause discolored and shrunken grain. Kernels blighted by scab disease will sicken hogs. The fungus bodies (sclerotia) produced on crops blighted by ergot are poisonous. The presence of blighted grains also increases the cost of cleaning and processing grain.

Often one organism or a similar one may incite the same or a similar disease on several of the grain crops. So I group the leaf blights and head blights according to the micro-organism that incites the disease: Bacterial leaf and head blights; fusarium head blights, or scab; ergot of grains and grasses; helminthosporium leaf blights and kernel blights; septoria leaf and culm blights; and rhynchosporium leaf and head blights.

Many of the bacterial leaf and kernel blights occur on cereals and grasses but are generally unimportant. They produce water-soaked stripes or halo-surrounded spots on the leaves. Leaf lesions frequently are abundant for short periods during the growing season, but they cause only minor damage unless they persist into the heading period.

Then the bacterial blights produce a sticky mass of bacteria around the spike. The result is a poor set of seed. Blasted heads are common when moist weather and bacterial blight occur during heading. Later kernel infection is marked by black or brown areas on the chaff and brown, sunken lesions near the germ end of the kernel. The bacteria also may be carried on the kernels without external evidence of their presence.

The bacteria that cause these diseases are minute, short, rod-shaped cells of various species and races. Some species are restricted to a single cereal or grass. Others comprise races, each of which may attack a given grain crop and its closely related grasses. Seed-borne bacteria are controlled by the use of organic mercury, but the bacteria that spread to grainfields from crop residues and grasses are not affected by seed treatments. Most of the adapted varieties of wheat, oats, and barley are not damaged greatly by bacterial blights. Very susceptible varieties, however, are unsuited to humid areas.

Fusarium head blight, or scab, a disease of wheat, barley, rye, and some grasses, sometimes causes extensive losses. It also damages corn and sorghum. Heavy losses are most frequent in the eastern and central Corn Belt and in similar humid or subhumid areas in this and other countries. The head blight develops during warm, humid weather at any stage from kernal formation to ripening of the grain. Infection occurs in the flowers and then spreads up and down the spike. The diseased part of the head soon turns a straw colour in wheat and rye and light brown in barley.

Frequently pink mold growth develops around the base of the flower and cements the chaff to the grain. The kernels of wheat and rye are shriveled, white or gray, and show a rough, scabby surface. In barley the kernels are light in weight and grayish brown; with the hulls removed, they look like scabbed kernels of wheat and rye. The fungus grows into the kernels and makes the starch floury, discolored, and partly changed to sugars. The proteins are partly broken down into soluble nitrogen compounds.

Some of the fats become rancid because of the formation of fatty acids. New compounds are produced that cause acute vomiting in dogs, pigs, and man, but do not affect sheep, cattle, or mature poultry. Grain containing 5 per cent or more of badly scabbed kernels, fed to pigs, causes vomiting, loss of appetite, and arrested growth. Scabbed kernels in grain therefore should not be processed into human food or used to feed swine. Scabbed kernels can be separated readily from the sound wheat, rye, or corn. Because it is hard to separate scabbed from healthy barley grain, scabby barley should be fed to cattle, sheep, or poultry.

Several species of Fusarium incite the disease on cereals and grasses. The most common species, F. graminearum, produces the sexual stage (perithecia) on old corn stalks and straw. The sexual stage is known as Gibberella zeae. The minute, blue-black spore cases (perithecia) of this stage are abundant on the surface of stalks and straw from April through June. The spores that incite head blight or ear rot are carried to the young heads of the small grains and to corn silks by the wind. Two other species, F. culmorum and F. avenaceum, are common north of the Corn Belt and in Canada. All of them also cause seedling blight and root and stalk rot in the cereals.

These pathogens are carried over on crop residues. Spores are not produced or dispersed if all crop residues are covered by a thin layer of soil. All straw and stalks should be plowed under and left below the soil surface to control the diseases. Surface mulching with straw and stalks invites head blight in small grains where wet, warm weather occurs after small grains are headed.

No highly resistant varieties of wheat, rye, or barley have been produced, but some varieties are more susceptible than others. All known varieties will be scabbed if spores are present and the weather is favourable from the heading period to maturity. Two barleys, Chevron (C. I. 1111) and Peatland (C. I. 5267), and three wheats, Haynes Blue-stem (C. I. 2874), Progress (C. I. 6902), and Rival (C. I. 11708), are among the least susceptible varieties. Investigations at the Minnesota and Wisconsin Agricultural Experiment Stations indicate that the inheritance of resistance is complicated.

Seed treatment of well-cleaned grain with organic mercury compounds will control the seed-borne infections. Ergot of grains and grasses is not strictly a head blight, although loss in yield from sterility is always associated with the disease. Resting bodies (sclerotia) of the fungus replace some of the kernels, and the adjacent kernels do not develop. The disease is common and destructive on grasses, rye, barley, durum wheat, and some varieties of hard spring wheat. The damage from it occurs in the humid sections and extends into the subhumid regions, notably in the spring grain areas of Nebraska, the Dakotas, and Montana.

The disease is recognised first by the sticky fluid (honeydew stage) on parts of the spikes soon after heading and later by the purplish-black sclerotia in the ripening heads. The honeydew attracts flies and other insects to the diseased heads. The sclerotia are shaped somewhat like a rye kernel. They usually are longer than a grain and thus protrude from the chaff. The sclerotia thresh out with the grain. The presence of ergot in flour or other grain products can be determined by standard colour tests. The ergot sclerotia contain several chemical compounds, some of which cause acute injury to animals and man. One group of similar compounds has special medicinal value. Bright, hard ergot sclerotia free from molds command a high price for medicinal use. Ergot sclerotia invaded by Fusarium contain compounds affecting the respiratory and other automatic muscular action.

Because most of the ergot sclerotia produced in the humid, central area are attacked by Fusarium and are unsuitable for medical extracts, supplies usually are obtained from the drier west central areas or imported. Ergot samples are tested before they are purchased by pharmaceutical companies. The ergot sclerotia contain compounds that cause contraction of the fine blood vessels. Investigations at the Montana Agricultural Experiment Station have confirmed early reports that milk flow in livestock is reduced greatly by continuous feeding of small quantities of ergot sclerotia. Relatively small quantities of ergot may cause serious losses among animals.

Any grain containing more than 0.3 per cent ergot sclerotia by weight is graded as ergoty and is discounted on the markets. The ergot content of milled products is limited by law. Although most of the sclerotia can be removed with modern cleaning machinery, it is difficult and expensive to bring the per centage down to the content permitted. The life cycle of the fungus, Claviceps purpurea, is well synchronised with that of the grain or grass host. The sclerotia falling on the soil or planted with the seed germinate when the grains and grasses are flowering.

The sexual spores are wind-borne to the flowers of the grain or grass, where they invade the young kernel and replace it with fungus growth. The first fungus growth forms a folded mat, which bears millions of spores in a sticky, sweet, honeydew-like mass. The minute spores are carried by insects or are spattered by rain to infect numerous other kernels. The sclerotia develop following this spreading stage. Some grasses are infected regularly from sclerotia dropped from the previous year's growth. The ergot spreads from the grasses to cereal crops, which are related to them botanically. Therefore, neither crop rotation nor ergot-free seed controls the disease.

Destroying grasses in the grainfields, particularly quackgrass, bromegrass, and wheatgrasses, and mowing grasses near grainfields before they head helps in control. Heavy pasturing, or mowing of grass pastures before heading, also reduces the danger of ergot poisoning of livestock. No varieties of barley, rye, or wheat resistant to ergot have been found. The Mississippi Agricultural Experiment Station has produced a paspalum grass resistant to the ergot that is common and destructive on this southern pasture grass. Helminthosporium leaf blights and kernel blights on cereals and grasses are caused by several species of Helminthosporium. Some of the diseases have been so damaging as to necessitate changes in varieties to maintain profitable grain production.

Helminthosporium blight, for example, forced much of the susceptible Wisconsin Barbless barley out of production in the North Central States not long ago. Another helminthosporium blight forced oats growers to drop varieties with Victoria parentage from production in the Central States. Such diseases are devastating when susceptible varieties are used throughout large areas.

This group of disease-producing fungi attacks seedlings, roots, crowns, leaves, and kernels. Spores, however, spread from the diseased seedlings to the leaves of healthy plants and from diseased kernels cause blighted seedlings. The spores, produced abundantly on crop residues, further spread the disease to the leaves and heads of growing grain. Seed treatment with organic mercury compounds, crop rotation, plowing under crop residues, and the use of resistant varieties are generally applicable control measures for this large group of cereal diseases.

Net blotch and kernel blight incited by Pyrenophora teres (Helminthosporium teres) occur on barley only. The characteristic symptoms are netted blotches or irregular stripes on the leaves, brown discoloration on the culm, and light brown, indistinct blotches on the kernels. The sise of kernels is reduced, but badly shriveled grain is uncommon. The sexual stage of the fungus is produced abundantly on the old stubble and straw. At least two distinct races of the pathogen have been identified.

They react differently on the two general groups of barley. C. W. Schaller, of the California "Agricultural Experiment Station, and R. G. Shands, of the Department of Agriculture and the Wisconsin Agricultural Experiment Station, found that resistance to the western race is concentrated in the Manchurian-Oderbrucker group of barleys, including Peatland, whereas resistance to the eastern race is concentrated in the North African and Abyssinian barleys. Only a few in the Manchurian-Oderbrucker group are resistant.

Stripe disease incited by Helminthosporium gramineum occurs on barley only. The sexual stage of this fungus is not found in nature and the life cycle differs from that of other species. Seedling infection results in fungus invasion of all plant parts, and spores are produced during the period of barley flowering. Seed infection occurs from wind-borne spores. All leaves of a diseased plant generally are affected. Chlorotic stripes appear on the young leaves as they unfold. The stripes turn brown and spread to the leaf sheath also when the leaves reach full development. All spikes of affected plants are blighted and brown in susceptible barleys of the Manchurian type. Diseased plants of the Coast barleys in California frequently develop some seed. Spores are abundant on the striped plants during and after flowering. Kernels infected from wind-borne spores do not show the disease in the dormant seed.

Several physiologic races of the pathogen occur, but practical resistance to the known races has been obtained and incorporated into commercial varieties. Inheritance of resistance is conditioned by two or more factor pairs with resistance dominant or partially dominant. Several genes for resistance are reported. Those used most extensively for breeding are from Lion, Peatland, Chevron, Brachytic and Persicum.

Leaf blotch and kernel blight incited by Pyrenophora avenae (Helminthosporium avenae) occurs on oats and several related grasses. The disease and its pathogen resemble those of net blotch of barley. Small, brown spots or irregular blotches, which may spread over the entire leaf blade, are the characteristic symptoms. Defoliation occurs in winter oats in the Southern States and on very susceptible spring oats in North Central States. Kernel infection is not conspicuous. Yields are reduced when the leaf blighting is severe.

Crop rotation, plowing under crop residues, seed treatment with organic mercury compounds, and using resistant varieties are the recommended control measures. Most of the commercial oat varieties are moderately resistant to the disease. Yellow leaf spot of wheat (Helminthosporium tritici-vulgaris) was reported first in Pennsylvania and Virginia, the eastern part of the region where soft red winter wheat is grown.

It has since spread westward into the hard red winter wheat area of Kansas and Nebraska. It is of minor importance. Leaf spot and blight of rye incited by Pyrenophora secalis (Helminthosporium secalis) first shows as a small brown spot, which spreads until the leaf blade is yellow and dry. Sexual spore cases are abundant on old rye straw. The disease is of minor importance if crop rotation and seed treatment are practiced. The first general occurrence of the disease was in North Central States in 1940.

Leaf spot (culm blight and kernel blight of barley, wheat, and grasses) is caused by Cochhobolus sativus (Helminthosporium sativum). The damaging seedling blight and root rot caused by the fungus is described. The brown leaf spot phase is common on barley. The spots spread to cause yellowing and the death of the leaves.

Diseased kernels of barley and wheat are dark brown or almost black, especially near the germ end of the kernel. That condition sometimes is called black point. Kernels infected early are shriveled and light in weight. The sexual stage of the fungus has been produced readily in culture through mating of compatible lines of the fungus by two Canadian scientists, R. D. Tinline and P. M. Simmonds.

The disease is controlled by maintaining good fertility of the soil, crop rotation, seed treatment with organic mercury compounds, and the use of resistant varieties.

D. C. Arny, of the Wisconsin Agricultural Experiment Station, found resistance in barley to be controlled by a single factor pair when using Lion as the susceptible parent and races of the pathogen found in the North Central States. The more resistant barleys are in the Manchurian group including Peatland and Chevron. Mars and Moore are resistant commercial varieties.

Victoria blight of oats, incited by Helminthosporium victoriae, a relatively new disease, has nearly eliminated several high-yielding, rust- and smut-resistant varieties that had been obtained from crosses with Victoria. It is primarily a seedling and culm disease. Leaves show a reddish-brown striping or spotting. The straw breaks over as the crop is heading.

The grain is shriveled and chaffy, largely because of infection at the base of the culm. The pathogen is carried over on the seed. The straw and stubble of susceptible varieties are darkened by the mass of conidia (asexual spores), which are produced at the nodes as the plants are killed and cover the surface of the straw in late fall and the following spring. Infected seed and crop residues spread the pathogen to the crop the following year.

Seed treatment with organic mercury compounds reduces seedling loss but does not control the later culm, blight. Apparently only oats derived from Victoria hybrids are very susceptible to the disease. Investigations at the Iowa and Wisconsin Agricultural Experiment Stations and at the Dominion Rust Laboratory, Winnipeg, Canada, indicate that susceptibility to Victoria blight is not completely linked with the Victoria type of crown rust resistance and that a factor for resistance to crown rust can be used without bringing in susceptibility to Victoria blight.

Septoria leaf and culm blights are common on cereals and grasses. The leaf blotches and spots frequently are indefinite in colour, shape, and margin. Usually the leaf and culm tissue surrounding the lesion bleaches slowly and turns straw colour or brown, depending somewhat on the host. As the leaves bleach or as the diseased plants approach maturity, the small, regular spore cases (pycnidia) develop beneath the leaf surface of the diseased areas. These are globe-shaped, with a definite opening to the surface and light golden brown to brown in colour. The abundant spores borne in these spore cases are slender to threadlike and colorless to light yellow. The different species are based largely on the length of the spores and on the hosts upon which they develop. The slender spores are wind-borne.

The damage they cause results largely from shriveled kernels and lodging. Shriveled kernels and reduced yields are common in susceptible varieties. The damage in susceptible varieties of wheat, barley, and oats is similar to that caused by stem rust.

Control is difficult as the spore cases are produced in abundance on the old straw and stubble. Crop rotation, plowing under old straw, and the use of resistant varieties offer the only means of control. Many of the older varieties are moderately resistant to the disease. Some of the newer varieties of wheat, oats, and barley are susceptible. Cereal breeders, in their search for better rust resistance, sometimes have used septoria-susceptible varieties as parents, but that has meant an increase in damage from this group of diseases.

The septoria blights of grains and grasses are incited by several different fungus species. Each is restricted to a particular grain crop and its closely related grasses. I list the common species on the grain crops. Wheat: Two species are common on wheat, Septoria tritici, a long-spored species that attacks mainly leaves and is widely distributed; and S. nodourum, a very short-spored species that attacks leaves, culms, and heads.

Barley: S. passerinii has thin, medium-long spores and occurs on cultivated barley and some of its wild relatives. Varieties of the Manchuria-Oderbrucker type are susceptible or moderately susceptible. Dorsett, Valentine and Jet used extensively in breeding are resistant. North African and Egyptian types often show considerable resistance. Atlas and Atlas 46 are resistant commercial varieties of the Coast type.

Oats: S. avenae has medium-narrow, short spores and produces the sexual stage abundantly on susceptible varieties. The varieties from Bond are relatively susceptible to culm rotting by this species, and perithecia develop on the old diseased straw. A relatively large number of grasses including wheat are attacked by this species and its varieties. Most of the commercial oat varieties other than those derived from Bond are relatively resistant. Rye: S. secalis has slender, medium-short spores. This species and variety occur on rye and related grasses especially in the north central area. Rhynchosporium leaf scald and kernel blight is primarily a leaf scald of barley, rye, and numerous grasses, although kernel blight is common in the southwestern barleys grown during the winter. The leaf scald appears as oval to irregular areas, which at first are water-soaked and gray green and later become zonated and yellow or brown and gray green. The pathogen produces a mat of tiny colorless asexual spores in the gray-green surface of the lesion. The spores are scattered by rain and wind. Spores are produced abundantly on crop residues especially during cool weather.

The pathogen has specialised races, which are restricted closely to the different small grain and grass species. Experiments at the California Agricultural Experiment Station indicate losses in yield of barley up to 20 per cent when leaf scald is severe. Early development of scald reduces the number of kernels; later infection reduces kernel weight. Atlas 46 some experimental varieties are highly resistant. Many of the Manchurian barleys are relatively resistant.

VIRUS DISEASES OF CEREAL CROPS

Most of the plant viruses impair or destroy chlorophyll, the all-important food-synthesising green pigment in crop plants. The plant struggles for its existence or dies if its chlorophyll has been impaired.

Some viruses stunt or otherwise deform or alter the growth of plants without seriously affecting the chlorophyll. Sometimes it seems that the chlorophyll even may be increased. Some plants may carry a virus but show no signs of disease.

Some viruses tend to impair or destroy the chlorophyll uniformly through the leaf. They cause chlorosis a general yellowing or bleaching. In others the impairment or destruction occurs in patches, spots, or streaks in the leaf; there is a light-green or yellow or white mosaic mottling, spotting, or streaking. The chlorotic patterns in the leaves are strikingly similar for many of the virus diseases of the grasses, even though the viruses are quite different.

Among the cereal crops in the United States the greatest losses from the virus diseases are caused by the mosaics of winter wheat. Several viruses attack cereals. Some have been controlled through the use of resistant varieties, but for a destructive group of viruses in the area between Oklahoma to South Dakota no effective control is known.

We identify the viruses of cereals and other plants largely on the basis of symptom reactions of the infected plants when grown under proper environmental conditions; the host-range characteristics; the methods by which a virus can be transmitted from diseased to healthy plants; and the ability of a virus to live in extracted plant juice or in dried tissue under different temperatures, and other physical and chemical treatments for given lengths of time. The method of perpetuating a virus is determined by the characteristics of the virus. If the virus can be transmitted only by tissue union or by insects, it is kept alive in the living plant or sometimes in the living insects. Many viruses that can be transmitted experimentally by manual methods of inoculation are kept alive for months or years in frozen juice or in dried tissues near freesing.

The wheat, barley, oat, and cucumber viruses that can be transmitted by manual methods of inoculation are kept in clipped leaf tissue that is dehydrated over calcium chloride, and stored over anhydrone at temperatures just above freesing. At intervals the stored viruses are increased in growing plants, and new dehydrated tissues are prepared and stored. The virus of brome mosaic and viruses of the Agropyron mosaic can be maintained in that way, but usually less work is involved when they are carried in their respective native host plants, and the new cultures are started by the division of the plant stools or from the rhizomes. The virus of barley stripe mosaic can be kept in infected seeds for at least 8 years.

Soil-borne wheat mosaic viruses occur in Illinois, Indiana, Iowa, Kansas, Maryland, Missouri, Nebraska, North Carolina, Oklahoma, South Carolina, and Virginia. Infected fields have been found in nearly one-half of the counties in Illinois. Indiana also has many infected areas. In many places, particularly in Illinois, the viruses would curtail the production of wheat were it not for resistant wheats.

These viruses of wheat and its close relatives also infect barley and rye, although they have shown relatively high resistance. Oats, corn, and plants outside of the grass family apparently are not susceptible to them. Bromus commutates, a wild grass, is susceptible. The viruses are not carried in the seed from the diseased plants.

The virus of wheat mosaic-rosette (Marmor tritici var. typicum), is the first virus that we knew could be carried over in the soil from season to season. The rosette phase of the disease attracted attention in 1919, near Granite City, Ill., but its virus nature was not recognised until 1925. Previously it was suspected that insects (particularly the hessian fly), cold winter temperatures, and several species of fungi were causing the trouble. In 1920 it was learned that the causal agent is associated with soils and that it can be inactivated in the soil by steaming and treatment with formaldehyde.

Many varieties of wheat were found to be immune to rosette. From 1920 to 1923 I selected rosette-immune strains from the highly susceptible varieties Harvest Queen and Illini Chief. Years later, when Illinois 2 wheat was found to be highly susceptible to mosaic rosette,

O. T. Bonnett and his associates at the University of Illinois selected lines that were immune to rosette and highly resistant to mosaic and that had several of the desirable characteristics of Illinois 2. Those lines contributed to the development of Prairie and Royal varieties.

The evidence that the causal agent Of rosette was associated with the soil actually delayed the discovery that a virus caused the disease. In 1920 workers generally considered that viruses did not over-season in the soil. Even today few viruses are known to be carried over in the soil. Fortunately the rosette symptoms occur in only a few varieties, such as Harvest Queen, Illini Chief, and Missouri Bluestem. Most varieties, however, develop mosaic mottling when they are grown in infested soils.

Along with the field experiments, microscopic studies that Sophia Eckerson, R. W. Webb, and I conducted at the University of Wisconsin revealed the presence of abnormal inclusion bodies in many of the cells of both rosetted and mosaic plants. These cell inclusions were so strikingly like those associated with some known virus diseases, and the mosaic symptoms were so evident in many of the varieties of wheat, that we strongly suspected the virus nature of rosette and the mosaics.

Inoculation methods commonly used to transmit mosaic viruses in plants like tobacco and cucumber failed to transmit the disease to healthy wheat seedlings. Neither rosette nor mosaic developed when I grew winter wheat out of its natural growing season. So I got the idea that cool temperatures favour infection and disease expression.

Tests by Dr. Webb demonstrated that soil temperatures near 60 F. and comparatively high soil moisture favoured natural infection from the soil. When I inoculated healthy wheat seedlings with the juice from the diseased plants and then cultured them at cool temperatures, they developed the disease.

Mosaic symptoms are transmitted most successfully by growing the inoculated plants with ample sunlight, a daily photo-period of 8 hours, and a temperature near 60. I observed that some of the wheat plants growing in the field developed light-green mosaic mottling. Others developed severe yellow mottling. In the inoculation tests with virus from mosaic spelt and wheat plants, I saw that some of the plants of Harvest Queen wheat developed light-green mosaic and rosette, while others developed yellow mosaic somewhat like the situation that I had observed in the study of the mosaic viruses that infect tobacco.

Virus-selection techniques soon enabled the isolation of two viruses. One induced mild light-green mosaic and rosette in Harvest Queen wheat. The other induced severe yellow mosaic in Harvest Queen, but no rosette. I found that most varieties of wheat developed only mosaic when they were inoculated with the mosaic-rosette virus. Some varieties developed light-green mosaic. Others developed mild yellow mosaic.

Some of the severe yellow mosaic viruses that have been isolated from the soil-borne series differ somewhat in their ability to destroy chlorophyll and to stunt or otherwise deform the plant. None, however, has been observed to cause the excessive stooling or bud proliferation caused by the mosaic-rosette virus. To distinguish them from other viruses that cause yellow mosaic in wheat, they are referred to as the prairie wheat yellow mosaic viruses (Marmor tritici var. fulvum). Attempts to isolate a virus that induces rosette without the mild mosaic mottling have failed.

These yellow mosaic viruses are regarded as strains of the mosaic-rosette virus, but further study is needed to determine whether they are direct mutants or more distant relatives. I have not seen the strong unilateral interference, characteristic of the tobacco mosaic virus in combination with its yellow mosaic virus mutants, in the soil-borne viruses isolated thus far from wheat. It is therefore harder to demonstrate mutation in the wheat mosaic viruses than it is in the tobacco mosaic virus.

The soil-borne wheat mosaic viruses are relatively unstable in extracted plant juices and in tissues removed from the plant and kept at ordinary temperatures. The viruses are inactive when the leaves die on the plant. When fresh mosaic leaves are dried in the laboratory, the virus becomes inactive within 2 or 3 weeks. Furthermore, the viruses soon die when fresh mosaic leaves are allowed to decompose in moist soil. The thermal death point of the yellow mosaic viruses tested thus far is 140 to 149 F., with 10 minutes exposure when in the plant juice. Traces of yellow mosaic virus passed a Schott-Jena fritted glass filter having average pore sise of 880 millimicrons. The dilution end points for the rosette and the yellow mosaic viruses in distilled water are between 100 and 1,000 times.

Mosaic may occur in an infested field every year for at least 12 years, or it may fail to appear after the first year. Air-dried soil stored in the laboratory for 3 years remained infectious. Some lightly infested soils lost their infecting power when stored through one summer, however.

In a cropping test that covered 5 years on infested land, Benjamin Koehler, of the Illinois Agricultural Experiment Station, found that continuous cropping with a susceptible wheat greatly favoured the occurrence of mosaic and rosette. Cropping for 4 years with an immune wheat, oats, soybeans, and corn greatly reduced or controlled completely the rosette and mosaic in the susceptible wheats grown on the land the fifth season. Four years of cropping with alfalfa also reduced the amount of disease, but less than other crops did. Under favourable conditions, which are not completely understood, Dr. Koehler found that a lightly infectious soil gradually became highly infectious after four successive croppings with susceptible wheats.

The effects of various chemicals and heat on infested soil have been tested by Koehler and me and also by Folke Johnson at Ohio State University. Infestation can be eliminated by the use of heat, formaldehyde, chloropicrin, dichloropropene-dichloropropane (D D), napthalene, calcium cyanide, carbon disulfide, methyl bromide, rotenone, napthalene, and ethyl alcohol.

Ethylene dichloride was less effective than the other chemicals. Ethyl chloride gave no control. Toluene gave no control in air-dried infested soil that was saturated with the chemical for 4 days, followed by complete evaporation of the chemical and seeding within 5 days to a susceptible wheat.

We do not know the exact relation of these wheat viruses to the soil. Perhaps they live in some soil-inhabiting organism that carries the virus and introduces it into the underground parts of the young seedling in autumn.

The Plains Wheat Mosaic viruses were collected by L. E. Melchers from a few scattered wheat plants growing in nurseries and fields near Manhattan, Kans., in 1931 and by me in 1941. Tests with the soil in which the mosaic plants were growing failed to indicate that the viruses are soil-borne. The viruses transmit by manual inoculation with some difficulty even at favourable temperatures of from 60 to 65 . They cannot be maintained in plants growing at summer temperatures at Beltsville.

We isolated green mosaic and yellow mosaic strains of virus. The green mosaic strain does not induce rosette in mosaic-rosette-susceptible Harvest Queen wheat. Unfortunately the viruses were lost in the summer of 1942, when no cool-temperature culture chambers were available.

Soil-borne viruses isolated recently from collections made in Kansas and Oklahoma are so similar to the plains wheat mosaic viruses that it is suspected that the viruses collected in 1931 and 1941 may have been soil-borne. The negative results obtained with the earlier collections of soil may have been due to a low level of infestation, and the plains wheat mosaic viruses may be strains of Marmor tritici.

Soil-borne viruses of oat mosaic occur in Alabama, Georgia, South Carolina, and North Carolina. The mosaics were observed first in the experiment station plots at Auburn, Ala. When they are abundant, they reduce the yields of grain and forage of susceptible oat varieties such as Bond, Camellia, Victoria, and Letoria. Growing resistant and tolerant varieties Anderson, Atlantic, Appler, Arlington, Custis, Fulgrain, Fulwin, Lee, Lemont, Victorgrain, Winter Resistant should prevent widespread losses from the mosaics. The mosaics have been observed in fields only in winter-grown oats and in oat relatives.

The viruses can be transmitted by manual methods of inoculation when the oat plants are grown at temperatures near 60 to 65 F. Two viruses have been isolated. One, Marmor terrestre var. typicum, causes apical mosaic, a light-green mosaic mottling that tends to be prominent towards the tip portion of the leaves of Letoria oats. The other, M. terrestre var. oculatum, causes eyespot mosaic, which in Letoria oats causes light-green or yellow, nearly elliptical spots, which have green centres. Some plants in certain oat strains that are grown in infested soil develop a typical rosette. A specific virus for oat rosette has not been isolated. The viruses of these mosaics are not carried in seed from diseased plants. In the soil they have shown much the same characteristics as the soil-borne wheat mosaic viruses.

Wheat Streak Mosaic viruses were first collected at Salina, Kans., in 1932, but records of L. E. Melchers make it seem likely that similar mosaics may have been present in Kansas before 1930. In 1922 G. L. Peltier found a wheat mosaic in Nebraska that had characteristics of streak mosaic.

The disease has been worst in western Kansas, but it has caused damage in Nebraska, Colorado, and South Dakota. Some damage has been reported in Wyoming and western Iowa. The disease is known to occur in Oklahoma, the Salt River Valley in Arizona, California, and Canada. In Nebraska and South Dakota spring wheats as well as winter wheats are affected.

Streak mosaics gradually increased over the years. In 1949 they caused an estimated loss of 7 per cent of the wheat crop in Kansas. That amounted to a loss of 30 million dollars. The loss in 1951 also was heavy. It is not uncommon to find entire fields and sometimes entire communities in which streak mosaic has wiped out the wheat crop. The destruction of the chlorophyll in the leaves frequently is so great that the plants are yellow.

Growth is halted and the plants fail to set seed. The viruses do not overseason in the soil. They are easily transmitted by manual inoculation at summer temperatures characteristic of Washington, D. C. Infection is erratic at high summer temperatures, such as those that prevail at Stillwater, Okla. They are not seed-borne. Maybe some insect carries them. The strains of streak mosaic virus range from those that induce severe yellow spotting, streaking, and mottling (Marmor virgatum var. opi. cum), to those that cause mild, light-green or weak-yellow streaking and mottling (M. virgatum var. viride). With the strains of virus studied thus far, chlorophyll destruction or yellowing is favoured by cool temperatures with the daily averages ranging from 60 to 70 . Wheat varieties differ in the degree of yellowing caused by a given strain of virus. Under certain conditions streak mosaics can be confused with the soil-borne virus mosaics.

The thermal death point of these viruses in plant juice is near 129.2 , with an exposure of 10 minutes. The survival time at room temperature in air-dried leaf tissue has not been more than 40 days. The dilution end point in water is near 5,000 times. The virus passed a Berkefeld "W" filter, pore sise 3 to 4 microns. Streak mosaic viruses occur in several wild grasses. They can infect corn, oats, barley, several species of winter annual bromegrasses, sand lovegrass, Canada bluegrass, some of the crabgrasses, and other range and pasture grasses. In some oats, the virus favours reddening in association with the chlorotic patterns. In Golden Giant sweet corn, streak mosaic viruses induce a rather typical mosaic mottling, rings, dots, dashes, and short streaks, some of which are elliptic and have green centres. The symptoms generally resemble those induced in field corn by the sugarcane mosaic virus.

The long, yellow, or bleached stripes associated with so many other viruses in corn are not characteristic of streak mosaic. The plants are never killed, and infection has never been obtained in all inoculated sweet corn plants. Some lines of sugarcane develop local lesions when the leaves of the true seedlings are inoculated with streak mosaic virus. The virus does not become systemic, though. Sorghum and the perennial smooth brome, Bromus inermis, are highly resistant or immune. Several of the susceptible forage and wild grass species may contain resistant varieties.

Tests have failed to reveal any plant outside of the grass family that is susceptible to the streak mosaic viruses. Because methods for inducing natural infection have not been worked out, the wheat variety test nurseries have to be inoculated by spraying. Varieties of wheat show some differences in the reaction to streak mosaic, but none has shown a safe degree of resistance to the severe yellow strains of the virus. High resistance and immunity have been found in rye, several wheat-grasses (species of Agropyron), and some of the hybrids between wheat and the Agropyron species. Barley Stripe Mosaic virus was identified with the barley false stripe disease in 1950, but the disease has been known since about 1910. It used to be thought to be of nonparasitic origin. False stripe was first noted in spring barley growing at the Wisconsin Agricultural Experiment Station at Madison. Pressed specimens of the diseased plants, prepared by A. G. Johnson on June 27, 1913, and still in existence, leave little doubt that the symptoms are the same as those induced by the virus. Perhaps it was the first virus disease on a grass species to be collected in the United States.

The disease occurs throughout the spring barley region in the United States, and it occurs in Canada. It has not been regarded as a menace to the crop. Pot tests out of doors prove, however, that it causes serious reduction in the yield of grain and forage in several varieties of spring barley. The brown stripes often induced by the virus in some plant species and varieties tend to distinguish it from other viruses associated with grasses. The most common symptoms associated with the virus, however, are the unmistakable mosaic types of chlorotic mottling, spotting, and streaking. The chlorotic areas in the leaves frequently are ashy gray to white because of the complete or nearly complete destruction of the carotinoids, or yellow pigments, as well as the chlorophyll.

The virus is transmitted easily by manual methods of inoculation at summer temperatures at Beltsville. It infects wheat, sweet corn, field corn, smooth crabgrass, and occasional seedlings of smooth brome. It has induced only local lesions in inoculated leaves of rice and tobacco. In Michigan Amber wheat seedlings, the first one or two leaves that develop systemic signs usually become almost solid ivory colour or white, denoting a marked, acute phase of the disease. The seedlings are not killed. The new leaves become much less chlorotic in the chronic phase. But when infection is through the seed, only the chronic symptoms appear, beginning with the first leaf of the wheat seedlings. In Golden Giant sweet corn and U. S. 13 field corn, the virus seldom kills the seedlings, but the plants become badly stunted and usually worthless.

The long yellow or bleached stripes in the foliage are like those induced by the brome mosaic virus, the cucumber mosaic viruses, and by some leafhopper- transmitted viruses that infect corn. With the small grains, cool temperatures favour strong symptoms in the infected plants. The thermal death point of the virus in plant juice is near 154.4 F., with 10 minutes exposure. The survival time in air-dried tissue at room temperature has not been more than 40 days. The dilution end point in water is slightly beyond 10,000 times. Traces of the virus have passed the Berkefeld "N" filter, but not the Berkefeld "W" filter. Outstanding is the ability of the virus to enter some of the seeds of the diseased barley and wheat plants. Very few viruses are seed-borne; this was the first to be discovered in a grass species. Because infected barley seeds tend to be poorly filled, many probably go with the screenings during threshing and thus tend to keep down the occurrence of the disease. Tests have failed to show the virus to overseason in the soil.

Wheat Striate Mosaic was discovered in winter wheat in South Dakota in 1950 by John T. Slykhuis, who succeeded in transmitting the virus to healthy wheat plants and reproducing the disease by means of the leafhopper Endria inimica. Attempts to transmit the virus by manual methods of inoculation have failed. The disease occurs also in Kansas and Nebraska. The extent of the losses caused by striate mosaic is not known.

In Minter and Rushmore varieties, the disease shows up in fine, light-green or yellow lines in a streaked pattern. In the early stages, the lines are in the tissues directly over the veins. In Minter, brown necrotic spots may develop later. In Rushmore and other varieties, the streaks are followed by premature yellowing and death of the leaves. The diseased plants frequently are so stunted they produce little or no seed. In the advanced stages, striate mosaic is readily confused with other yellow mosaics of wheat. The host range of the virus is not known.

Barley yellow dwarf, another new virus disease of cereals, was widespread and destructive in California in 1951. John W. Oswald and Byron R. Houston found the virus to be transmitted by five widely distributed species of aphids, the corn aphid (Rhopalosiphum maidis), the apple grain aphid (R. prunifoliae), the English grain aphid (Macrosiphum granarium), the grass aphid (M. dirhodum), and the greenbug (Toxoptera graminum). Attempts to transmit this virus by manual methods of inoculation have failed.

The chlorotic symptoms induced by the virus are similar to those caused by nonparasitic factors an excess of soil water, drought, a shortage of nitrogen, and low-temperature injury to the chlorophyll system. The light-green or yellow mottling, the broken streaking, and the spotting that characterise the other virus diseases of the small grains are relatively temporary or absent in yellow dwarf. Leaves of infected plants rapidly turn light green and yellow, beginning at the tips. The chlorotic foliage tends to redden in oats, as when some of the non-parasitic factors are operating.

Diagnosis therefore is difficult, except in fields where the nonparasitic factors and other diseases can be excluded from consideration. The most certain diagnosis is through transmission tests with one or more of the aphid vectors. Plants become infected at all stages. Young plants are frequently killed. Stunting gradually decreases with advancing age of the plant. Only the flag leaf shows signs of yellowing or reddening when infection takes place in the late stages of development. As with other virus diseases, the yield of grain shows the greatest reduction when the plants are infected in their early growth. The wheat variety Kanred has shown some tolerance to yellow dwarf, but inoculated plants were stunted.

Agropyron or quackgrass mosaic viruses were collected at Arlington Farm, Va. The green mosaic virus (Marmor agropyri var. typicum), was collected in 1934, and the yellow mosaic virus (M. agropyri var. flat,zim), was collected in 1936 in the same area from which the green mosaic virus was collected. Mosaic was found in 1950 in quackgrass in South Dakota by John Slykhuis and in Iowa by Vernon E. Wilson. In 1951 mosaic quackgrass was found on the Plant Industry Station grounds at Beltsville. At Arlington Farm, the green mosaic virus was isolated once from wheat growing near the patch of mosaic quackgrass.

The viruses collected at Arlington Farm transmit to quackgrass and to wheat by manual inoculation, but with some difficulty. Infection and symptoms are favoured by temperatures ranging from 60 to 65 , but infection will take place at summer temperatures. At the higher temperatures, however, the green mosaic frequently becomes masked and the yellow mosaic becomes very mild. The viruses overseason in the rhizomes. Tests with soil in which the mosaic plants were growing in the field failed to indicate over-seasoning in the soil.

All attempts to infect Agropyron repens with the soil-borne wheat mosaic viruses have failed. Corn stunt was found in 1945 in the San Joaquin Valley, Calif., by N. W. Frazier, and in the lower Rio Grande Valley, Tex., by George E. Alstatt. The studies of L. O. Kunkel showed that corn stunt virus is transmitted by the leafhopper Dalbulus (Baldulus) maidus. The virus infects many varieties of field corn, sweet corn, and teosinte, a close relative of corn. Attempts to infect the small grains, other grasses, and nongrass species have been unsuccessful. The virus cannot be transmitted from diseased to healthy plants by manual methods of inoculation, but Karl Maramorosch, of the Rockefeller Institute for Medical Research, succeeded in transmitting the virus to leafhoppers by manual means.

The disease was reported from the Mesa Central, Mexico, by J. S. Niederhauser and J. Cervantes in 1950. They found another leafhopper, Dalbulus (Baldulus) climatus, that transmits the virus. In the early stages of corn stunt, the chlorotic spotting, streaking, and banding in the leaves resemble the markings associated with some of the mosaics. As the disease progresses, the signs resemble those associated with the corn streak disease. In the advanced stages the leaves may be completely yellow or bleached; the leaves and stalks also may show a tinge of red. chlorotic markings appear in the sheaths and husks. The upper part of the stem may become chlorotic. The tassels may be deformed.

Infected plants tend to have short internodes and extra shoots that originate from the axillary buds and from the ear branches, so that the plant is stunted and bushy. Husks tend to have unusually long tips. The ears produce few kernels. The roots and the brace roots sometimes branch profusely and are stunted. The disease tends to be more extensive in late corn than in early corn, probably because the insect carriers are more abundant late in the season. The virus is not carried in the seed from diseased plants.

Sugarcane mosaic is the first virus disease of grasses to be identified as such. The virus, Marmor sacchari, was introduced from the Tropics. Soon after its discovery, E. W. Brandes found out that the virus is transmitted by the corn aphid, and that the virus infects corn, sorghum, pearlmillet, and many other grasses in the Southern States where cane is grown. The virus apparently does not infect any of the small grains. Several strains of the virus can be differentiated on sugarcane.

At least four species of aphids transmit the virus. In studies by Hugo Stoneberg, this mosaic had no apparent effect on the rate of growth or the total height of the corn plants, and tended to increase suckering and the number of ears slightly. Yields were reduced less than 10 per cent, and the quality of the corn was reduced slightly by the disease. The southern celery mosaic virus, a strain of cucumber mosaic virus (Marmor cucumeris), infects corn in several parts of Florida. Natural infection is by the melon aphid, Aphis gossypii. Infections have been induced in teosinte, field corn, sweet corn, kafir, sorgo, milo, wheat, and rye by means of the aphid in studies conducted by F. L. Wellman.

In my tests with this virus and related cucumber mosaic viruses, I was unable to infect wheat from manual inoculations, although infection could thus be induced in Golden Giant sweet corn. In it the yellow stripes in the leaves are like those induced in corn by the mosaic (stripe) virus occurring in Cuba, Trinidad, and Hawaii; by the barley stripe mosaic virus; and by the brome mosaic virus when the corn plants are inoculated after the seedling stage to avoid early death. The cucumber mosaic viruses frequently kill sweet corn seedlings and young plants in greenhouse tests. Diseased plants that survive early infection are worthless for crop production. The Brome Mosaic virus (Marmor graminis) has been collected in nature only on smooth brome (Bromus inermis) from three locations near Manhattan, Kans., and Lincoln, Nebr. No insect vector for it is known. The virus induces local or systemic signs in a wide range of grass species, including wheat, barley, oats, rye, corn, sorghum, and many wild grasses. Local lesions develop on the inoculated leaves of some breeding lines of sugarcane in the young seedling stage. It is one of the few viruses that can infect certain nongrass species, besides grass. Its range among the nongrass plants seems to be more limited than Pierce's disease virus of grape, southern celery mosaic virus, and certain other strains of the cucumber mosaic group, which also infect a few grass species.

Brome mosaic virus induces local lesions in the wiped leaves of Samsun Turkish tobacco, Early White Spine cucumber, Scotia beans, some varieties of sugar beet, garden beets, Swiss chard, and lambs quarters (Chenopodium album). In lambs quarters all inoculated plants give local lesions, and some of the plants in a population develop systemic chlorotic spotting.

In the small grains, the symptoms it induces can be confused with those of the wheat yellow streak mosaic virus. The virus kills young seedlings of Golden Giant sweet corn, although the seedlings gain resistance as they grow. When older plants become infected, their growing points are usually killed, and the foliage develops yellow stripes or streaks resembling those induced by the barley stripe mosaic virus, the cucumber mosaic viruses, and several of the leafhopper-transmitted viruses that infect corn.

To determine the spread of brome mosaic in the field, close plantings in small plats of alternated healthy and mosaic smooth brome and orchard-grass plants were observed by C. L. Lefebvre and me. We started the test in the spring of 1946 and continued it through the 1947 crop season. During the period of the test, the plants were often walked on to simulate conditions in a pasture. At the end of the test, 12 out of the original 23 healthy bromegrass plants and 6 out of the original 27 healthy orchardgrass plants had become infected. The disease reduced the forage yield about90 per cent in the smooth brome and about 80 per cent in the orchardgrass.

Of the viruses that have host ranges chiefly among the grasses, the brome mosaic virus has the highest thermal death point for 10 minutes), the longest survival time in air-dried tissue at room temperatures (more than 20 months), and the highest dilution characteristic (100,000 to 300,000 times in water). The virus passed a

Schott-Jena fritted glass filter having an average pore sise of 780 millimicrons. The Virus Of Pierce's Disease of grape and of alfalfa dwarf (Morsus suffodiens), infects an exceedingly wide range of plant types, including grasses and a sedge. Eight species of grass are known to become infected in nature in California. One is wild oats. Fifteen species of grass have been tested experimentally, and 12, including barley, became infected with the virus.

Other small grains were not tested An outstanding characteristic is the inability of the virus to induce apparent signs of disease in most of the plants it infects. None of the infected grasses has expressed symptoms, yet these hosts carry large amounts of the virus, and they are favoured by the leafhoppers that transmit the virus. The association of the virus with the xylem tissue may account for the suppression of symptoms in most plant species. The fact that symptoms occur in grape and in certain legume species, however, suggests the possibility that symptoms might be incited in other plants by providing suitable environmental conditions.

Several virus diseases of cereals have not been found in the United States. All of the known viruses infecting cereal species outside of continental United States and Canada, except the soil-borne wheat mosaic viruses in Japan, are transmitted by insects belonging to some one of the so-called hopper groups of the order Homoptera. None is transmitted by manual methods of inoculation.

Corn streak virus and wheat stunt virus (Fractilinea maidis), cause serious diseases in corn and wheat in parts of Africa. The best control of corn streak obtains when there is a break in the succession of crops, as corn seems to be the best carry-over reservoir of the virus. Three species of leafhopper, Cicadulina mbila, C. zeae, and C. nicholsi, transmit the virus. The first symptoms in corn consist of small, almost circular, colorless spots. As the disease progresses, the new leaves develop narrow broken streaks, which are yellow and sometimes translucent.

The "A-form" of the virus species causes the wheat stunt disease in South Africa. The disease is destructive in early sown spring varieties because the warm conditions favour the leafhopper vector, C. mbila. Resistant and immune wheats are known. The chlorotic spotting and streaking in the leaves suggests some of the yellow mosaics of wheat. The infected leaves tend to be curled. The bunchy, stunted plants resemble plants that have the wheat mosaic-rosette disease.

Corn mosaic of Hawaii and the corn stripe disease of Cuba are caused by the same virus, Fractilinea zeae. The virus has been reported also from Trinidad, Tanganyika, and Mauritius. It may occur in Puerto

Rico. It has been found in sorghum. Transmission of the virus is by means of the leafhopper Peregrines maidis. In corn, the chlorotic markings on the foliage resemble those associated with corn streak, as well as those associated with some of the mosaic viruses that can be transmitted by manual methods of inoculation. Corn wallaby-ear virus (Galla zeae), occurs in Queensland, Australia.

It is transmitted by the leafhopper Cicadula bimaculata. In young corn plants the virus induces elongated swellings, or galls, on the secondary veins on the under side of the leaves. The leaves tend to roll inward, as under conditions of drought, and their green colour is intensified. The plants are dwarfed. When older plants become infected, the reactions tend to be mild. Russian winter wheat mosaic virus (Fractilinea tritici) is limited to Russia, as far as we know. It is transmitted by the leafhopper Deltocephalus striates. The virus infects winter and spring wheats, rye, barley, and oats. It induces light-green or yellow mottling and streaking in the leaves. Infected plants may be dwarfed. Some varieties of winter wheat proliferate, causing rosette. Infected young plants are sometimes killed. The virus is not soil-borne.

Pupation disease virus of oats (Fractilinea avenue) is also known as zakooklivanie and Siberian oat mosaic virus. It apparently occurs only in Siberia. The virus is transmitted by the leafhopper Delphacodes (Delfax) striatella. It infects oats, wheat, rye, barley, rice, corn, and several wild grasses. The symptoms in oats resemble those described for wheat mosaic caused by the virus Fractilinea tritici. The virus of rice stripe disease in Japan is also transmitted by Delphacodes striatella, suggesting that the virus F. avenue may be the causal agent of this rice disease.

Rice dwarf virus (Fractilinea oryzae) occurs in Japan and the Philippines. The disease is also called rice stunt. The virus infects rice, wheat, rye, oats, and several wild grasses. It does not infect barley, corn, or sorghum. Infected rice plants are dwarfed. The young leaves have light-green or yellow spots along the veins. With the succession of new leaves, the markings consist of chlorotic spots and streaks. The virus is transmitted by at least two species of leafhoppers, Nephotettix apicalis (bipunctatus) var. cincticeps and Deltocephalus dorsalis. Experiments with N. apicalis have shown that the virus passes through part of the eggs to the next generation, for as many as seven generations.

THE SMUTS OF WHEAT, OATS, BARLEY

Many millions of dollars' worth of grain are destroyed every year by the smuts of wheat, oats, and barley. For purposes of study and control, we can consider the smuts as being seedling-infecting or floral-infecting. The seedling-infecting species come in contact with the host plants as follows: The microscopic spores from smutted plants are carried by wind, rain, insects, and other agencies to the heads of healthy plants (as in loose smut of oats). Or, smutted heads are crushed in threshing and spores are distributed to the clean seed or blown to fields, where later they come in contact with the host at seeding time (as in the stinking smuts of wheat).

As the seed germinates and the seedling grows through the soil to the surface of the ground, the smut inoculum develops thin threads, which penetrate the seedling and initiate infection. The fungus then grows internally in the plant and eventually forms spores in the young heads. That completes the cycle and sets the stage for a new crop of smut. Stinking smut, or bunt, apparently has been a plague of wheat since wheat was first cultivated. It was prominent among the diseases studied by the earliest plant scientists.

It has been widely investigated, and certain control measures have long been known. Nevertheless bunt is still an economic threat to the production of wheat, especially winter wheat, in all important wheat regions of the world. The average annual loss in the United States is estimated to be 1.3 per cent of the crop, or about 25 million dollars. Its severity fluctuates from year to year and from region to region.

The heaviest losses occur in the winter wheat regions of the Midwest and Pacific Northwest. Estimated annual losses in Kansas over 30 years range from less than 200,000 dollars to almost 20 million dollars; the average is almost 3 million dollars. In recent years the Pacific Northwest has suffered the most. The loss there was about 10 million dollars in 1950. The main loss comes from the reduction in yield. The per centage of reduction in yield roughly equals the per centage of smutted heads in the field.

Losses in quality result when market grain is discounted in price because of the dark colour and the offensive odour of the smut spores that adhere to the kernels. Smut that is removed from the grain by cleaning and washing is assessed as dockage against smutty grain. Stinking smut increases the cost of processing. It also is a fire hazard to threshing equipment and storage bins. Two kinds of bunt occur on wheat. The common bunt is caused by Tilletia caries and T. foetida. Dwarf bunt is caused by T. caries.

Common bunt is more prevalent and better known than dwarf bunt. Dwarf bunt stunts infected plants severely. It has been recognised as a distinct type only since 1925. Its prevalence and severity have increased steadily. Its principal region of distribution is the Pacific

Northwest, but it also occurs in Wyoming, Colorado, and New York. Wheat plants infected with bunt are recognised first by their reduced height. Common bunt shortens the plants a few inches or as much as half the height of healthy plants, depending on the physiologic race of the bunt fungus, the host variety, and the conditions of growth. Plants with dwarf bunt are one-half to one-fourth the height of healthy plants. Because of intergrading types, identification solely on the basis of stunting is sometimes difficult.

The smutted wheat heads are bluish green when they emerge from the boot. The healthy heads are yellowish green. Smutted wheat heads tend to be long and lax and to ripen sooner than healthy ones. The smut balls protrude beyond the glumes as they enlarge. Dwarf bunt infected heads usually are more compact than those infected with common bunt, and the glumes are spread apart so that the smutted heads have a feathery look.

The smut balls vary considerably. The sise depends on host variety, location in the spikelet, physiologic race, and climate. Common bunt balls are elongate or round. Dwarf bunt balls are always round. Common bunt balls are more fragile and absorb water less readily than those of dwarf bunt. Dwarf bunt stimulates excessive tillering of infected plants. The effect is more pronounced in the Turkey variety than in others. Forty tillers on dwarf bunt infected plants are not uncommon.

Bunt-infected flowers have longer pistils and longer and broader ovaries than do healthy flowers. Diseased ovaries are green; healthy ones are white. Stamens in diseased flowers are reduced in length and breadth, and the anthers have a pale-yellow colour instead of a pronounced green, as in healthy ones. The stamens in diseased heads fail to extrude at flowering time. Bunted plants are more susceptible to seedling blights and to yellow stripe rust but more resistant to powdery mildew than are healthy plants. Winter injury is more pronounced in bunted plants. Increased sensitivity of infected plants to gravity has been reported.

The spores of Tilletia caries are uniformly globose and have "netted" spore walls. Spores of T. foetida are globose, elongate, or oval and have smooth walls. The bunt balls of T. foetida are larger and more elongate than those of T. caries on the same variety of wheat. In both, the ball shape tends to conform to the shape of the wheat kernel. The spore balls are broken in threshing. The grain becomes contaminated with spores. If it is used for seed, the spores germinate and penetrate the young seedling in the course of its subterranean emergence from the seed to the surface of the soil and infection is established in the growing point. The growth of the parasite keeps pace with plant development; at maturity, bunt balls are formed in place of wheat kernels. Wheat seedlings also may become infected by soil-borne spores. In less humid regions like the Pacific Northwest, airborne spores from the combine harvester settle on summer fallow land and remain dormant until moisture and temperature are favourable for germination in the fall.

Usually that occurs at the time winter wheat is seeded, so that smut spores and wheat seeds germinate at the same time, thus exposing the seedlings to infection. Spores of common bunt perish in the soil in less than a year. Those of dwarf bunt may remain viable for 7 years. Even so, dwarf bunt does not attack spring wheat. The greatest infection occurs at soil temperatures of 40 to 60 F., with moisture content ranging from 15 to 60 per cent of field carrying capacity. Other factors affecting the development of bunt are soil fertility, depth of seeding, and length of day.

The spore load is important. About 0.5 gram of spores to 100 grams of seed are necessary for maximum infection. The greater the concentration of spore load around the embryo, the higher the degree of infection. Both species of the bunt fungus are highly specialised into distinct physiologic races. The races differ in their ability to attack different varieties, also in the sise and shape of their smut balls and rate and mode of spore. germination. In the dwarf bunt race infection comes primarily from soil-borne inoculum.

Pathogenic specialisation in the bunt fungi was discovered in 1925 or so. Different races have been identified from various parts of the world. In the United States about 25 races are recognised by their reaction to a set of eight differential wheat varieties. New or previously unrecognised races frequently appear, especially in places where varieties resistant to the recognised races have been introduced. The new races attack and spread with the new varieties. Every smut-resistant variety introduced in the Pacific Northwest eventually has become the distributor of one or more new races of bunt.

The two species as well as different races of the bunt fungus have been intercrossed artificially and new pathogenic lines were selected from the hybrid populations. The frequent occurrence of intermediate spore forms in field collections indicates that natural hybridisation also occurs. Invariably the intermediate forms are found on highly susceptible varieties. These varieties, however, tend to perpetuate the old, established races, despite their susceptibility to the new races.

Effective control of bunt depends on the use of clean seed of a smut-resistant variety properly treated with an appropriate fungicide.

Also effective, when it is practical, is the seeding of wheat when soil temperatures are unfavourable to bunt development. Those control measures seem so simple that they give no hint of the long and costly struggle of plant scientists to develop them. And the struggle continues today in an effort to keep pace with the ever-changing problem, brought on by shifts in wheat varieties and by the adaptive nature of the smut fungi.

The accidental discovery of seed treatment as a means of controlling wheat smut was made long before the parasitic nature of the disease was known. Some wheat seed salvaged from a grounded sailing ship off the southern coast of England was sown by nearby farmers. The crop they got had less smut than that from local seed. That may have been the origin of the 100-year-old practice of brining the wheat seed. A scientific approach to the problem of smut control by seed treatment began about 1800 with the discovery of spore germination and the harmful effect of copper sulfate on their germination. Fifty years later the parasitic nature of wheat smut was established, and the quest for more effective control by seed treatment was begun.

Scientists recognised at the outset that a good chemical for treating seed should be highly toxic to the smut spores but harmless to the seed. Copper sulfate did not always give good control and often injured the seed. Seed had to be soaked in the solution and dried afterwards. It left much to be desired as a suitable fungicide for seed treatment. It took almost 50 years of search to find a better treatment, but progress was faster thereafter. The discovery of the value of formaldehyde about 1895 was hailed as the answer to smut control by seed treatment. After its adoption in the United States, formaldehyde rapidly replaced copper sulfate because of its several advantages.

Even so, it still had the disadvantage of being a wet treatment and was likely to injure the seed unless used properly. The impelling need for a dry treatment led to the discovery of the value of copper carbonate dust in Australia. Copper carbonate was first tested in the United States about 1918. It gradually displaced formaldehyde for the control of wheat smut but was ineffective against oat and barley smuts, for which formaldehyde continued to be used.

Paradoxically, the elimination of the wet feature of the formaldehyde and copper sulfate introduced the dust hazard of the dry treatments. Many workers were made ill when they inhaled the dusts. Efforts to eliminate this undesirable feature brought the development of the slurry treatment, by which the fungicide is applied in a concentrated water suspension or slurry in such small amounts that the seed is not wetted appreciably. It avoids the main disadvantages of both wet and dry treatments. It became the predominant method of treating seed wheat in the Pacific Northwest within 5 years.

The value of resistant varieties in the fight against wheat smut is illustrated by experience in the Pacific Northwest, where soil contamination occurs and where seed treatment is practiced universally. Between 1931 and 1942, when resistant varieties predominated, the incidence of wheat smut declined from more than 30 per cent to less than 3 per cent. In the succeeding years up to 1951, when susceptible varieties predominated, the incidence of smut again increased to 30 per cent.

That different varieties of wheat reacted differently to smut was observed almost 200 years ago. Only 50 years ago did plant breeders begin to breed smut-resistant varieties systematically. At first they thought that complete control of bunt could be realised through the use of resistant varieties. They emphasised the development of varieties with smut resistance.

The first of the new varieties carried the recommendation that they could be grown without seed treatment. Invariably, however, they became susceptible to new races of smut. Breeding for smut resistance consequently is recognised as a perpetual contest between the directed processes of man to produce resistant varieties of wheat and the processes of nature to develop new races of bunt. The breeding of smut-resistant varieties has been facilitated by the systematic identification of different genetic factors for resistance.

At least two major and numerous minor factors are recognised. Probably others exist. Thus far, however, the two major factors control the entire gamut of physiologic races now recognised, but this simple situation may not always remain. Consequently it is advisable to develop additional varieties with maximum resistance to smut while continuing to search for new sources of resistances.

The best temperature for bunt infection is 37 to 57 F. Seeding early (before the soil temperature drops that low) or late (after it has fallen below that point) tends to reduce the amount of smut. Both early and very late seeding, however, hurt the wheat crop. A reasonable delay in seeding should aid in control where soil contamination prevails, as in the Pacific Northwest.

Seeding practices have never been widely adopted as a means of smut control; they have obvious limitations, particularly that sowing must be done when the soil is moist enough to germinate the wheat. Flag smut occurs on wheat in most of the world's major wheat countries but has caused heavy losses only in Australia. Its presence in the United States was established first in 1919 in Illinois. It was found in Missouri and Kansas shortly thereafter. The source of infection was never discovered, but imports of wheat for feed from Australia were suspected. Considerable apprehension followed the discovery of flag smut in the United States.

Quarantines were established to prevent further importation of wheat from Australia and to prevent the spread of the smut outside the known infested area. Those measures or natural factors may have operated against the spread of the disease; flag smut was not found outside the original area of infestation until 1941, when it was reported in central Washington. Up to 1953 it had not become widely prevalent or intensively severe in any locality, but its persistence in the major wheat areas posed a constant threat.

Infected plants usually are dwarfed and have twisted leaf blades and sheaths marked with grayish-black stripes. The stems also may be invaded. Infected plants rarely produce heads, but smutted heads occasionally appear. At maturity the diseased tissues dry up and become shredded. The stripes rupture and expose the black spore masses. Symptoms may appear on infected plants at any time from the fourth leaf stage, or 1 month after germination, up to the heading stage One or all culms of a plant may be infected. The oldest culms of partly smutted plants are smut-free. Partial smutting is more common in the greenhouse than in the field and in some varieties more than in others. Infected plants may die and dry prematurely, thus obscuring much of the loss from the disease.

Flag smut of wheat is caused by the fungus Urocystis tritici. Its spores occur singly or in balls, which usually are covered by a layer of sterile cells and contain three spores. Individual spores are smooth, oval to spherical, dark brown, and 12 to 16 microns by 9 to 12 microns in sise. Germinated spores have a short promycelium on which 2 to 6 sporidia are borne. The sporidia remain attached to the promycelium, fuse, and give rise to infecting mycelium. Infection occurs in the seedling stage from spores carried on the seed or in the soil. Systemic development inside the plant follows.

The cycle is complete with the production of spore masses in the diseased tissues. The number of races of U. tritici is relatively small. Five races have been identified in China. Only two races are known in the United States. One occurs in the Midwest. The other, more widely virulent, is in the Pacific Northwest. Germination of spores and infection take place in a wide range of temperature and humidity. Spores germinate best at temperatures of 40 to 80 F. Soil temperatures of 57 to 70 are favourable for infection. The highest infection occurs at 66 to 70 . Optimum moisture content apparently shifts with prevailing temperatures. With soil moisture at 40 per cent, best infection occurs at 50 to 59 . At 60 per cent soil-moisture infection is best at 50 . At temperatures above 59 , best infection occurs when soil moisture is below 40 per cent. Spore viability is best retained at relative humidities of 50 to 75 per cent.

Seed treatment with a suitable fungicide destroys spores carried on the seed. The use of resistant varieties prevents infection by soil-borne spores. Because the spores of flag smut may remain viable in the soil more than one season, crop rotation helps control the smut. Smuts on oats seem to be among the first cereal diseases to come under the scrutiny of early writers on plant diseases. Their prevalence and economic importance are' world-wide. They destroy 40 to 50 million bushels of oats in the United States each year. Total losses aggregating more than 27 million bushels were reported in 1945 from Iowa, Minnesota, Illinois, and Wisconsin, the leaders in the production of oats. The heaviest infections occur in the South Atlantic region, where losses up to 35 per cent have been reported in some States.

The least damage occurs in the West. The two kinds of smuts on oats destroy the panicles by producing masses of spores in place of oat kernels. Loose smut, caused by Ushlogo avenae, destroys the enveloping glumes as well. The spores, borne in a loose mass, are quickly disseminated by wind currents, leaving behind a denuded panicle skeleton.

Covered smut, caused by U. kolleri, produces spores in balls enclosed by the glumes. The branches are shortened, so that the panicle has a compact appearance. Covered smut panicles ripen prematurely, but the spore balls remain intact until broken up in harvesting. A buff-colored type of covered smut was collected from experimental material, but it has not been observed in commercial fields of oats.

Scientists in experiments have demonstrated hybridisation between species and races of the smuts and the segregation and recombination of pathogenic and other characters to produce intergrading types. The occurrence of intergrading types in the field is evidence that hybridisation occurs in nature. New types of smut are produced in that way.

Both oat smuts infect the seedlings, develop systemically in the host, and finally destroy the oat kernel. Spores of loose smut are disseminated by wind immediately after the diseased panicle emerges from the leaf sheath. Infection thus is spread to flowers of healthy panicles in the same field or other fields. The spores germinate in the flowers and produce mycelium, which establishes itself between the developing kernel and the glumes or chaff. The mycelium becomes dormant as the kernel matures. The smut mycelium becomes active again when the oats germinate and the seedling is infected.

The covered smut, on the other hand, emerges from the host with the spores enclosed in the glumes. The spore balls remain intact until broken in harvesting and threshing. The spores are then distributed over the surface of the threshed seed. They remain dormant until the seed germinates, when they also germinate and infect the seedling. The dissemination of the loose smut spores by wind makes it ten times more prevalent than covered smut, which spreads only by contact with infested seed.

Covered smut is rare in the South Atlantic region. Loose smut occurs least often in the western region. The degree of infection by the smuts is determined chiefly by soil temperature and moisture when the seedlings emerge. Infection may occur at temperatures between 41 and 86 F., but usually it is highest between 59 and 68 . Infection may occur when soils have a moisture content between 5 per cent and 60 per cent, but chiefly at about 35 to 40 per cent. Soil reaction near the neutral point or slightly acid seems to favour infection. Response to environmental factors is highly variable, possibly because the smuts have such a high degree of specialisation.

About 25 races of the oat smut fungus species are recognised in the United States by the reaction of 10 host testers of the species Avena sativa and A. byzantina. All the major commercial varieties grown in 1953 were susceptible to one or more races of smut. The Victoria oat was highly smut-resistant for at least 15 years and was used as a source of smut resistance in breeding new varieties. Races of smut that attack varieties derived from Victoria are widely prevalent throughout the southern half of the United States.

That experience with the formerly smut-resistant Victoria oat emphasises the perpetual nature of the problem. Smuts on oats can be controlled by growing resistant varieties and by seed treatment positive measures that should eliminate most of the loss if they are applied diligently everywhere. But seed treatment is not universal, and smut continues to reduce the crop. New smut races that attack formerly resistant varieties appear at times. The covered smut of barley is found the world over. In the United States it destroys half a million to 4 million bushels annually.

The disease first becomes noticeable at heading time, when smutted heads emerge from the boot. The hard, black masses of smut along the axis of the affected heads are covered with a grayish membrane. Each smutted head contains millions of microscopic spores. In experimental cultures a good infection is obtained by inoculating the seed with a suspension of spores. The seed is washed and dried to loosen the hulls around the kernel, the suspension is mixed with the grain, and the moist inoculated seed is stored for 16 to 20 hours to promote spore germination and the spread of mycelium under the hull.

Higher infection has been obtained in plants kept under mild temperatures for 2 to 4 weeks after the seedlings emerged than when the plants were grown continuously under cool conditions like those of fall and winter. During the initial infection, between seed germination and the emergence of the seedling, the highest per centages of smutted plants are obtained at soil temperatures of 50 to 70 F. The amounts of smut tend to be higher in plants grown in acid soil than in neutral or alkaline soil. Soil type, soil compaction, depth of seeding, and the rate of seedling growth also may affect development of smut. Thirteen physiologic races of the barley covered smut fungus have been isolated by their reactions on eight varieties of spring barley. One prevails in the winter barley region, another in the Mississippi Valley, and a third race along the Pacific coast.

Several varieties have shown high resistance to covered smut. Two varieties, Ogalitsu and Jet, are resistant to 13 races of covered smut, 9 races of nigra loose smut, and 6 races of nuda loose smut. Covered smut may be controlled by treating the seed with formaldehyde or organic mercury dusts. Nigra Loose Smut of barley, Ustilago nigra, was first discovered in 1932 by V. F. Tapke, who separated it from ordinary loose smut. The two loose smuts, which often were mixed together in a field, previously had been regarded as a single type that infected barley only through the flowers.

The two loose smuts look alike, but the nigra pathogen is a seedling-infecting smut and is controlled by treating the seed with surface disinfectants. When the two loose smuts occur together only partial control can be obtained by surface disinfection because the deep-borne loose smut is not affected by such treatment. The fungi causing barley covered smut and nigra loose smut have been hybridised artificially. Nigra loose smut causes an average annual loss of 1 million bushels of barley in the United States.

The disease usually first becomes noticeable at heading time, when dusty, smutted heads appear. Each contains millions of loosely held, dark-brown microscopic spores. Wind carries the spores. Spores that come in contact with the flowers and young developing seeds of healthy heads then behave like the spores of covered smut. They may lie dormant or germinate and form a sub-hull mycelium. Moisture, temperature, and other conditions affect germination of spores. Temperatures of 60 to 70 and a relatively dry soil are most favourable to infection of the seedling during the period of emergence.

Immediately after the seedlings have emerged, temperatures of 60 to 70 for 10 to 30 days give higher per centages of smut than do constant low temperatures. Nine distinct pathogenic strains or physiologic races of the nigra loose smut fungus from this country have been isolated by their reaction on eight varieties of spring barley. Four are rather widespread in the United States. At least four races that do not occur in the United States have been found in smutted barley from various parts of Israel.

At least four varieties of barley are highly resistant to all races of the pathogen thus far found in the United States. None is grown commercially, but all are useful in breeding productive resistant varieties. Only Two of the small grain smuts are floral-infecting. The smutted heads appear at heading time. They shed their spores during the bloom period of the normal heads. The loosely held spores are distributed by wind, rain, insects, or other agents.

Initial infection rarely occurs after the fertilised ovary has attained one-third of its mature sise. Spores that come in contact with healthy flowers germinate by forming infection threads, which grow down the pistil or through the ovary wall into the young, developing seeds. A deep infection of the seed follows, so that surface seed disinfectants do little good. The fungi also are unable to produce infection when spores are applied to the surface of ripe seed.

The two floral-infecting loose smuts, one of wheat and one of barley, are practically identical except that one attacks wheat but not barley and the other attacks barley but not wheat. Both are widely distributed in humid and subhumid areas but are less common in dry areas. Infection threads develop too slowly in dry air to penetrate the ovary during its brief period of susceptibility. Wheat loose smut caused annual losses ranging from 3 million to 18 million bushels in the United States from 1917 to 1939. Loose smut in barley caused annual losses of 750,000 to 4.5 million bushels in those years.

High humidity favours infection by loose smut in wheat and barley, but the ranges of temperature and humidity most favourable to infection have not been determined exactly. Certain wheats that rarely develop loose smut when grown under the dry conditions of western United States are extremely susceptible when grown in humid climates. The experiments in the Netherlands show that wheat loose smut spores may be disseminated and cause infection up to at least 100 yards from their point of origin. Infection diminished regularly with distance from the source of inoculum where high winds prevailed during the flowering season; it was spotty when air currents were irregular. A warm soil when seedlings emerge seems more conducive to smut than does a cold soil. In Japan, for example, early seedings of winter wheats and barleys often show higher amounts of loose smut than do later seedings. Grains that follow rice show little smut because seeding then is delayed for at least 6 weeks.

Soil fertility seems to have little influence on infection. Wheat plants grown from infected seed are more susceptible to winter-killing, especially under severe conditions, than are similar healthy plants. The deep-infecting loose smuts of wheat and barley can be controlled by hot-water seed treatment. The usual method involves immersing the seed 4 to 12 hours in unheated water, followed by a 10-minute dip in water at 129 F. A single immersion in water held at 120 for 95 minutes also is effective in control.

Such treatments are so laborious that individual growers rarely use them. The treated seed is soft, swollen, and hard to dry before seeding. The treatment often injures germination. Occasionally central treating plants, established where steam or hot water are available, treat the seed for local growers. Control also has been obtained by soaking the seed for 6 hours in water at room temperature followed by a soak of 40 hours in a 0.2-per cent solution of Spergon.

FOUR ENEMIES OF SORGHUM CROPS

The sorghum crops include grain sorghum, sorgo (or sweet sorghum), broomcorn, Sudangrass, and Johnson-grass. Four general types of diseases attack them: Those that rot the seed or kill the seedlings; those that attack the leaves and lower the value of the plants for forage; those that attack only the heads and so prevent the normal formation of grain; and those that cause root or stalk rots and prevent the normal development of the plant.

Seed rot is most severe when the soil is cold and wet after planting a common condition in the North and in other areas when seed is planted early. Much of the seed then fails to germinate and rots because it is attacked by various seed-borne and soil-inhabiting fungi. To germinate promptly, sorghum seed requires a relatively warm soil, above 70 F.

Most seed-rotting fungi thrive at lower temperatures that retard the germination of the seed and give the harmful fungi an opportunity to attack it. Some fungi (chiefly species of Fusarium, Aspergillus, Rhizopus, Rhizoctonia, a, Penicillium, and Helminthos-porium) invade and destroy the endosperm, the starchy tissue of the seed, thus robbing it of the food necessary to produce a strong seedling. Cracks in the seed coats of the kernels give the fungi ready access to the interior and thus aggravate the trouble.

Some fungi, especially species of pythium, attack the young sprout in its early stage and prevent its emergence. These fungi also attack and rot the primary roots and thus keep the young seedlings from getting enough food materials from the soil to become well established. Fusarium moniliforme, besides rotting the seed, frequently attacks sorghum seedlings at the surface of the soil soon after they have emerged and causes them to rot or damp-off and fall over. It also may destroy the primary roots of young seedlings. Some races of the fungus are more harmful than others.

Penicillium oxalicum attacks the endosperm and thus arrests germination. It may also kill the seedlings even after they have reached the third- or fourth-leaf stage. Seed rot and seedling blight may be controlled pretty well by careful selection and treatment of seed and proper cultural practices. Seed should be well matured and properly dried. The seed coat should be as free as possible from cracks and nicks, such as those that are often caused by improper adjustment of the threshing machine. Before being planted, the seed should be treated with a good disinfectant that will protect it from seed-borne fungi and, to a great extent, from the harmful fungi in the soil. The seed should be planted after the soil is warm enough for prompt germination.

Sorghum leaf diseases may range in severity from small, unimportant Spots or stripes on the leaves to diseased areas covering practically the entire leaf. High temperatures and humid weather generally favour the leaf diseases. The diseased spots or stripes are usually discolored because of chemical substances or pigments that are produced in the plant cells whenever they are injured. In most varieties of sorghum and Sudangrass this pigment ranges from reddish or brownish purple to almost black.

In most broomcorn, kaoliangs, and a few other sorghums, the spots or stripes on the leaves are red. In shallu, Ellis sorgo, and Tift and Sweet Sudangrass they are tan. Certain environmental conditions or hereditary factors occasionally discolor leaves of sorghum and Sudangrass. Those spots often are confused with symptoms produced by fungus or bacterial diseases.

A common condition in sorghum is the presence of intensely colored leaf spots or stripes without any other indication of disease. Much of this nonparasitic spotting may be due to mechanical injuries from insect punctures, wind, or sand particles. When chlorate weed-killing chemicals or grasshopper bait containing arsenic fall on the leaves of sorghum, they cause a burning effect in irregular but characteristic spots, which resemble those caused by parasitic leaf diseases. Often, however, the cause of this leaf spotting apparently is a physiological breakdown of the leaf tissues. Occasional plants have leaves so badly discolored that most of their leaf area is involved. The spots may be solid, or they may follow various concentric or irregular patterns. Some of the latter types are hereditary.

Leaf diseases may be caused by bacteria or fungi. Those caused by bacteria usually are characterised by the presence of drops or films of exudate that dry to thin, crustlike scales. Leaf spots caused by fungi have no exudate and usually are more or less roughened, because of the presence of fungal fruiting bodies.

Bacterial leaf diseases occur in the United States wherever sorghum is grown. Like most leaf diseases, they are favoured by warm (75 to 85 F.), moist weather. The causal organisms may be carried over from one season to another on the seed, on infected plant material in or on the soil, and occasionally on plants that overwinter. They may be spread from one leaf or plant to another by wind, splashing rain, and insects. Infection takes place through breathing pores of leaves.

The bacterial diseases usually do not cause serious losses, because they generally do not develop fully until the plants have reached their full sise. During warm, moist seasons, however, they may spread rapidly from the lower to the upper leaves until one-half to two-thirds of the leaf surface is destroyed. The forage value of the crop is impaired and the kernels may not fill properly.

Three bacterial diseases of sorghum are known in the United States: Bacterial stripe, bacterial streak, and bacterial spot.

Bacterial stripe, caused by Pseudomonas andropogonz, the most serious and abundant, attacks grain, forage, and sweet sorghums, broomcorn, and Sudangrass.

Bacterial streak, caused by Xanthomonas holcicola, occurs on the leaves of sorghum and Johnsongrass as narrow, water-soaked, translucent streaks about one-eighth inch wide and 1 to 6 inches long.

Bacterial spot, caused by Xanthomonas syringae, attacks the leaves of sorghum, broomcorn, Sudangrass, Johnsongrass, pearl millet, foxtail millet, and corn. On sorghum the spots appear first on the lower leaves; infection gradually spreads to the upper leaves as the plants approach maturity. Frequently the spots are so numerous that they unite into large diseased areas and kill the whole leaf.

Recommended control measures are sanitation, seed treatment, and the use of resistant varieties. Disposing of old infected plant litter and infected plants that overwinter, along with crop rotation, will reduce the quantity of inoculum present in the fields the next season. Seed treatment before planting will keep the disease from being carried over on the seed. The sorgos as a class seem to be more susceptible to bacterial stripe than are grain sorghums and Sudangrass. The kafirs are resistant to bacterial streak.

Eight distinct fungus leaf diseases, caused by as many different fungi, are commonly found on sorghums in the United States: Rough spot, anthracnose, leaf blight, zonate leaf spot, gray leaf spot, target spot, sooty stripe, and rust. The rough spot disease, caused by Ascochyta sorghina is widespread in Florida, Georgia, Alabama, South Carolina, North Carolina, Louisiana, and Mississippi.

It attacks sorghum, Sudangrass, and Johnsongrass. It is first observed as circular or oblong, light-colored spots. Then usually the red or tan pigment, depending on the variety, becomes apparent as the fungus spreads and injures the leaf tissue. Soon small black specks, the young fruiting bodies of the fungus, develop in the injured spots. The most striking characteristic of rough spot is the abundant development of these fruiting bodies (pycnidia), usually on the surface of the diseased discolored area, but occasionally on green, healthy-appearing parts of the leaf surface.

When the affected areas are rubbed between the fingertips, the sandpaper-like roughness, caused by the hard, raised fruiting bodies, can be detected readily. By the time the leaves die and become dry and papery, the fruiting bodies often are so abundant that they cover most of the leaf surface. Similar lesions occur on the leaf sheaths and occasionally on the stalks.

Sorghum or Sudangrass should not be grown on land where rough spot occurred the preceding season. Seed treatment and the use of available resistant varieties are advisable. Anthracnose occurs commonly on the leaves of sorghum, Sudangrass, Johnsongrass, and other grasses grown in the humid areas of the South. Most varieties of broomcorn are especially susceptible.

The disease is caused by the fungus Colletotrichum graminicola. It may be carried on the seed and also may live on dead and decaying plant refuse on or in the soil. Infection often causes spots to develop on the leaves when the plants are still in the seedling stage. Later the disease may spread to other leaves as they appear.

Usually, however, the leaves are not affected severely until about the middle of the growing season, when the plants have reached the jointing stage. Infection first appears on the leaves as small, circular or elliptical spots, which later enlarge and may unite to involve large areas of the leaf. The leaf midrib, which is commonly infected along with the leaf blade, is often strikingly discolored. Later the centres of the leaf spots fade to a grayish-tan colour; examination with a hand lens reveals the presence of numerous pin-point black specks with short, stiff hairs. Those are the fruiting bodies of the fungus, which, under moist conditions, produce pinkish spore masses. The spores are spread by rain and wind to other leaves, where they start new areas of infection.

Defoliation due to anthracnose reduces the value of the plants for forage and may reduce the sugar content of the stalks in very susceptible varieties. It also may lower the ratio of sucrose to invert sugars. Clean culture and rotation to avoid planting sorghum in fields cropped the previous year to Sudangrass, sorghum, or Johnsongrass should reduce the losses due to anthracnose. The principal means of control is growing resistant varieties.

Leaf blight, caused by Helminthosporium turcicum, is most prevalent in the warmer humid Atlantic and Gulf Coastal Plains of the Southern and Southeastern States, where it causes serious losses in sorghum and Sudan-grass. It also attacks corn. It probably is the most destructive Sudangrass disease of this area. A greenish, moldlike growth of spores develops in the centre of the leaf spots during warm, humid weather. The spores are scattered by wind or rain and infect other leaves. When the weather is favourable, the disease spreads rapidly and may cause serious damage by killing parts or all of the leaves before the plant has matured.

The chief hope of controlling leaf blight lies in the development of resistant varieties. All commercial strains or lots of Sudangrass that have been tested are highly susceptible. Plant breeders are attempting to develop resistant strains of Sudangrass from crosses between Sudangrass and resistant varieties of sorghum. Tift Sudan shows some resistance but is not immune.

Rotation does not appear to be an effective method of control, because the fungus lives in the soil for several years. Seed treatment may prevent some seedling infection and spread of the disease to new areas. Zonate leaf spot, caused by Gloeocercospora sorghi, attacks sorghum, Sudan-grass, Johnsongrass, and also sugarcane, corn, and cattail, or pearl, millet. It has been observed in eight Southeastern States. The disease is conspicuous on sorghum leaves as reddish-purple bands of tissue, alternating with tan or straw-colored areas and forming a zonate pattern.

Not much is known about the damage caused by this disease. When plants are so heavily infected that the leaves are killed prematurely, the forage value of the crop is undoubtedly reduced. No fully proved control measures are known. The fungus has been found in the glumes and seed, which suggests that the planting of disease-free or adequately treated seed would help to prevent the spread of the disease. No highly resistant varieties are known.

Gray leaf spot, caused by Cercospora sorghi, occurs on sorghum, Sudangrass, Johnsongrass, and corn. It is not certain that the races that attack corn are the same as those that attack sorghum. The disease is of minor importance, but occasionally it causes considerable spotting of sorghum leaves in the more humid and warmer sections, particularly in the Gulf States. It is commonly found on Johnsongrass along highways and fences.

The development of resistant varieties appears to be the most feasible control measure. The reaction of sorghum varieties to this disease, however, is not yet known. Target spot, caused by Helminthosporium sorghicola, occurs on sorghum, Sudangrass, and Johnsongrass. It has been found in five States. It produces small, well-defined spots, which are tan on Tift Sudangrass and certain other Leoti crosses and reddish purple on common Sudangrass and other varieties of sorghum. Older lesions have light centres surrounded by alternate dark and light bands; they look like a target.

The fungus fruits sparsely under ordinary field conditions; hence its spread is not rapid. Spores form in abundance in a moist atmosphere. The use of resistant varieties seems to offer the likeliest means of control, but none has yet been developed. As the disease may be seed-borne, effective seed treatment will prevent the spread of the disease to new areas in infected seed. Sooty stripe, caused by the fungus Ramulispora sorghi, occurs on sorghum, Sudangrass, and Johnsongrass. It has been observed in 10 States and on many varieties of sorghum and Sudan-grass and Johnsongrass.

It attacks the leaves and sheaths. On the leaves, the spots begin as small, oblong, reddish-purple areas, with purplish borders and straw-colored dead centres. These dead centres usually are more or less densely covered with small black bodies (sclerotic of the fungus), which may impart a sooty appearance to the lesions hence the common name of the disease, sooty stripe. In some varieties, particularly Leoti sorgo and Tift and Sweet Sudan-grasses, the borders around the leaf stripes are tan instead of purple. In Johnsongrass the purple border usually is less pronounced than in most varieties of sorghum.

The use of resistant varieties offers the most promising means of control, but none has yet been developed. Sorghum rust, caused by Puccinia purpurea, attacks Sudangrass, Johnson-grass, and most varieties of sorghum. It occurs frequently in the Gulf coast region and occasionally during wet seasons in States as far north as Nebraska and Indiana. Usually it does not become evident until the seed is well developed, so that it causes relatively slight losses to the grain sorghum crop except in certain seasons, such as that of 1950. Abundant rust, however, causes the leaves to dry and break off so that the forage value of the crop may be lowered. The method by which sorghum rust is carried over winter is not known, but the abundance of Johnsongrass throughout the South suggests that it might serve as the principal overwintering host for the fungus.

Growing resistant varieties is the only feasible method for controlling sorghum rust. The smuts of sorghum in the United States are covered kernel smut, loose kernel smut, and head smut. Covered kernel smut is caused by a fungus (Sphacelotheca sorghi) that attacks all groups of sorghum, including Johnsongrass. It is practically the only disease in which injury to the plant is confined almost entirely to the head. It occurs wherever sorghum is grown, but it is most prevalent in the Kansas-Oklahoma-Texas area.

Usually all, but occasionally only a part of the kernels on a smutted plant are affected. In smutted heads, enlarged cylindrical or cone-shaped smut galls are formed instead of the kernels. At first the smut galls are covered with a light-gray or brown membrane that later may break and release the dark-brown spores. Threshing breaks up the galls and spreads the spores to the healthy seeds.

When smutted seed is planted, the spores germinate along with the seed. The growing fungus then invades the developing seedling and continues to grow undetected inside the plant until after heading, when the smut galls, which have formed in place of the kernels, become evident. Plants affected by covered kernel smut appear normal except for the smutted heads.

Covered kernel smut can be effectively controlled by properly treating the seed, planting only smut-free seed, or growing resistant varieties. Because it is not safe to assume that seed is entirely free from smut and because resistant varieties of all types of sorghum are not available, seed treatment is the most logical means of control.

Loose kernel smut, caused by the fungus Sphacelotheca cruenta, is much less common than covered kernel Smut. It attacks all groups of sorghum, including Sudangrass and Johnson-- grass, although certain varieties in some groups are immune or highly resistant.

The galls formed by loose kernel smut are long and pointed. The thin membrane over them usually breaks soon after the galls reach full sise. Most of the dark-brown spores are soon blown away, leaving a long, dark, Pointed, curved structure, called a columella, in the central part of what was the gall. As in covered kernel smut, the spores of the fungus are carried on the seed and germinate soon after the seed is planted, when the fungus invades the young sorghum plant. It continues to grow unobserved inside the plant until after heading, When the long pointed smut galls appear in the heads in place of normal kernels. Unlike covered kernel smut, however, this disease stunts the infected plants and frequently induces the development of abundant side branches.

Loose kernel smut, besides being seed-borne and able to infect sorghum seedlings, may cause secondary infection that is, the spores from a smutted head may infect and cause smut to develop in late heads on otherwise healthy plants.

The control measures for loose kernel smut are the same as those for covered kernel smut; namely, seed treatment and the use of smut-free seed and resistant varieties. The treatments that control covered kernel smut will also control loose kernel smut.

Johnsongrass frequently is infected in the Southwestern States with a peculiar type of loose kernel smut caused by Sphacelotheca holci, which differs in several respects from that commonly found on sorghum. This smut also attacks some varieties of sorghum. Several feteritas and feterita crosses are susceptible. The kafirs appear to be immune, although they are highly susceptible to the other kernel smuts.

Infected plants head early and are severely stunted. Often they get to be less than a foot high and sometimes die prematurely. The thin membrane covering the smut gall ruptures as soon as the gall appears, and the spores are spread at an early stage. The spores under ordinary conditions lose their viability within a few months. In Johnsongrass the smut is transmitted largely through the underground rhizomes and also by spores falling on freshly cut stubble. Second-growth sorghum also may be infected through the stubble.

The control of this smut on its natural host is not important because Johnsongrass mostly is a noxious weed. Because of its short-lived spores, this smut is not considered a menace to sorghum. Head smut, caused by the fungus Sphacelotheca reiliana, attacks sorghum, Sudangrass, and, to some extent, corn. It is not common in the United States, but occasionally is somewhat damaging in individual fields of some varieties of sorghum.

Although head smut has been known in this country since 1890, the total losses from it have been small. Head smut is distinguishable from the kernel smuts because it destroys the entire head, transforming it into a large mass of dark-brown, powdery chlamydospores. The smut first becomes evident at heading time, when the large gall bulges out of the boot. The gall is covered at first with a whitish membrane, which soon breaks and allows the spores to be scattered by the wind and rain to the soil and to plant refuse, where they over-winter. The following spring and summer the spores germinate in the soil and produce sporidia, which, in turn, infect the sorghum plants. After invading a young sorghum plant, the fungus grows within it until the plant reaches the heading stage, when the smut gall becomes evident.

Because this smut fungus is carried in the soil, sorghum grown from clean seed planted on infested soil may be attacked. Some of the smut spores from the broken galls also may contaminate the seeds produced on nearby healthy plants. When such infested seeds are sown, head smut may be introduced into the soil of previously noninfested fields.

The most serious root and stalk diseases of sorghum are periconia root rot (milo disease), weak neck, and stalk rot. Except for periconia root rot, which under some conditions may appear at a relatively early stage, these diseases usually do not become evident until the plants are almost mature. Periconia root rot (milo disease) is caused by the fungus Periconia circinala and has been especially destructive to milo. It was discovered in 1925 in Texas, and in Kansas on irrigated land that had been cropped to milo for several years. Since then the disease has been observed in Oklahoma, New Mexico, Nebraska, Arizona, and California.

The disease apparently does not damage sorghum on any land not cropped previously to susceptible varieties. It is the most serious disease known on milo, darso, and their hybrids and, until resistant varieties became available, was a limiting factor in the growing of these varieties in infested areas. Other sorghums, with few exceptions, are not affected. In heavily infested soil, the disease may appear 3 to 4 weeks after planting, when the plants are only 6 to 9 inches high. The first indication of the disease is a stunting of the plants and a slight rolling of the leaves; the older leaves turn light yellow at the tips and margins.

This yellowing and drying progresses until all the leaves are affected and the plants die, usually without heading. They look as if they had been injured by excessive drought, alkali, or chinch bugs. In less heavily infested soil, the disease may not appear until the plants are about ready to head. Then it progresses less rapidly; the plants may grow weakly until late in the season and may form small, poorly filled heads. The disease attacks the roots before the above-ground parts of the plants show any symptoms. When affected plants are only a few inches tall, examination reveals a water-soaked brown or reddish discoloration of the outer part of the roots. A soft rot later destroys most or all of the fine root system and the outer part of the larger roots, while the central part of the large roots turns dark red or brown. The tissue at the base of the crown also turns dark red, and this discoloration extends up into the base of the stalk.

After the disease appears in a field, it becomes more severe each year that susceptible varieties are grown. It may appear at first in a few isolated spots in which the plants are stunted or retarded or may have died prematurely. The following year these areas will be larger and the plants growing in them will be more severely affected. If a susceptible variety is grown a third year, the entire crop may be completely destroyed early in the season.

The disease may be spread by soil or carried in run-off or irrigation water and by farm implements, wind-blown soil, or any agency that transports soil from infested fields. Although small areas of badly infested soil can be sterilised effectively by steam, formaldehyde, chloropicrin, or other agents, it is not economically feasible or practicable to do so in larger fields.

Effective control measures are limited to the planting of resistant varieties. Highly resistant strains of all desirable but hitherto susceptible varieties of milo have been developed. They have largely replaced the susceptible varieties. The combine-type grain sorghum varieties now widely grown are resistant to the disease. Crop rotation offers small hope for controlling periconia root rot because the causal fungus persists in the soil for several years.

Weak neck has become a serious farm problem in some areas since the introduction of combine harvesting of grain sorghums. The principal and most objectionable feature of weak neck is the breaking over of the peduncles the upper part of the stalks so that the heads fall to the ground and are missed by the combine. Weak neck is the result of over-ripeness accompanied by an inherent weakness of the tissues in the rachis (the centre stem of the head) and the peduncle, especially of the main stalk, of certain dwarf varieties of sorghum.

Before we had the combine, grain sorghums usually were harvested soon after the grain was ripe and while the peduncle was still moist and rigid. Now, however, they often are combined long after the grain has matured or after a freeze, when the upper part of the stalk has lost its sap and the grain is dry enough for safe storage. Under those conditions some dwarf varieties frequently break over at the base of the peduncle, which by this time has become dry and spongy so that in wet weather it absorbs moisture readily, becomes limp, and is easily broken over by the wind and the weight of the head.

Since weak neck is largely a varietal characteristic, the remedy lies in developing and growing combine types of grain sorghum having peduncles like those of sorgo, which remain green for a considerable period after the grain is ripe. Two such varieties Westland and Midland are now widely grown. Another possibility of avoiding loss lies in the development of combine varieties that have a stalk with a stiff rind, which will support the head after the peduncle has dried out.

Stalk rot of sorghum has become increasingly important since it was first generally observed in 1938. Most of the stalk rot and lodging used to be attributed to the fungus causing charcoal rot, but now it seems that several other fungi may be involved. Some of them apparently invade the plant through openings caused by insects or by mechanical injuries. Bacteria also invade the stalk and thus help bring about a water-soaked, and later, a rotted condition. There are four fungi to which stalk rot has been attributed. While no one of them is definitely known to be the sole cause, each may have a part. The diseases believed to be caused by them are known as charcoal rot, fusarium stalk rot, colletotrichum stalk rot, and rhizoctonia stalk rot.

Stalk rot may follow a period of drought, extreme heat, or other unfavourable conditions that weaken the plant. The disease is favoured also by injuries to the stem, crown, or roots caused by cultivation implements, insects, wind, and hail, or any other agency that makes an opening for the entrance of destructive fungi and bacteria. Definite methods for control of the four stalk rots are not known, although resistant varieties offer the chief hope of reducing losses caused by them. Some varieties appear to be resistant to charcoal rot and to colletotrichum stalk rot, but none has been observed to be resistant to the other stalk rots.

Rotation and other cultural practices may prove helpful, as will also the control of insects that attack the stalks of sorghum plants and leave openings through which stalk rot fungi gain ready entrance. Charcoal rot, caused by the fungus Macrophomina phaseoli (or Sclerotium bataticola), is the most destructive of the stalk rots. It is unpredictable and more or less sporadic in its appearance. It is associated with crop sequence and soil and weather conditions that subject the crop to extreme heat or drought at a critical stage in its development.

Usually injury is not apparent before the plants approach maturity, when many poorly filled heads are evident along with lightweight kernels and premature ripening and drying of the stalks, many of which are lodged. The diseased stalks may be soft and discolored at the base, the pith is disintegrated, and the separated vascular fibres have a shredded appearance. After a period of dry, warm weather, the fibres become covered with small black sclerotia formed by the fungus. When the roots and stubble decay in the field, the sclerotia become incorporated with the soil where they may germinate later and infect the roots of any one of 30 or more different crops.

Another soil-borne fungus, Rhizoctonia solani, which attacks cotton and several other crops, also may cause a stalk rot of sorghum. This stalk rot differs from charcoal rot in that it first attacks the pith, giving it a reddish colour, while the vascular fibres remain white. Large brown sclerotia are formed on the outside of the stalk.

The colletotrichum stalk rot, which is severe on broomcorn, is caused by the same organism, colletotrichum graminicola, that causes anthracnose on the leaves. The stalk rot phase usually is preceded by the anthracnose stage. The fungus enters the stalk directly through the rind and spreads rapidly throughout the conducting tubes and vessels of the plant. That interferes with the movement of water and food material and results in inferior heads and seeds. Diseased stalks frequently break over at the base or at a point several joints above the ground, making harvesting difficult.

That and the poor quality of broomcorn brush mean that many fields of broomcorn remain unharvested. Control measures are based on clean culture, crop rotation, and the use of resistant varieties. Treatment of sorghum seed every year with an effective fungicide is a cheap form of crop insurance. Treatment of sufficient seed for most farms is economical with regard to chemicals and labour, because only 2 to 5 pounds is usually required to plant an acre. This quantity can be treated at a cost of a cent or less for material.

Fungi, which may rot the seed or kill the seedling, can be combated to a great extent by treating the seed with an effective fungicide, of which there are several on the market. The fungicides also control the two kernel smuts and prevent the spread of head smut to the soil of other fields by means of spores on the seed. Although seed treatments cannot be depended upon to prevent bacterial and fungus leaf diseases, they may prevent their spread to new areas.

Chapter 6

Vegetable Crops

FIELD DISEASES OF BEANS AND LIMA BEANS

More than 50 diseases attack beans and lima beans in this country. In a season when weather favours their development, they may cause losses to American farmers of 15 million dollars. No region of the United States is free of all bean diseases. Root rots cause serious trouble wherever crop rotations are not practiced. Frequent rains during the growing season are likely to lead to heavy losses from bacterial blights, especially if seed that carries the blight organisms is planted.

Bean rust often causes serious losses in parts of the South and West. Virus diseases occur in all bean-growing districts. Downy mildew of lima beans often gives considerable trouble in the North Atlantic States. About 30 years ago, when anthracnose and bacterial blights were causing heavy losses of snap beans, research revealed that the organisms causing those diseases, which are seed-borne, do not thrive in areas where low rainfall and high temperatures prevail during the growing season.

Consequently the business of growing seed of snap beans moved from the East and Middle West to the drier intermountain region. Seed growers in parts of that region and in the Pacific States need have little fear of large losses from the two diseases, and crops grown from their seed all over the United States are less apt to be diseased than those grown from seed produced east of the Rockies.

Another line of research that has enabled farmers to grow healthier bean crops is the development of fungicides for control of certain diseases. Still another is the development of better systems of crop rotation and better methods of field sanitation. Possibly the most progressive step of all has been the development of disease-resistant varieties. A number of new varieties, each resistant to several diseases, have been introduced.

The Bacterial Blights are among the most important diseases of beans wherever the crop is subject to frequent rains during the growing season. If occasional showers occur during the early part of the growing season the diseases are usually noted even though the total rainfall is not excessive. Only in certain sections of the Intermountain and Pacific Coast States, such as southern Idaho, eastern Washington, and California, are beans grown relatively free of these diseases. Practically all beans grown for seed purposes therefore are produced in a rather small area in the West. The severity of the diseases in any place varies with the weather and the presence of infected seeds.

The bacterial blights include common blight, fuscous blight, halo blight, and bacterial wilt. Each is caused by a different bacterial organism, but all, except the wilt organism, produce similar symptoms on bean plants. Under certain conditions the bacterial wilt organism produces a wilting of the plant unlike that caused by any of the other organisms. Under other conditions some of the wilt symptoms are like those of other bacterial diseases.

Water-soaked spots on the leaves usually are the first visible evidence of blight. The spots enlarge, turn brown, and finally kill the leaf. Halo blight makes a halolike zone of greenish-yellow tissue around each water-soaked spot in cool weather. Leaves of newly infected plants are yellow. Leaves of plants infected with the common or the fuscous blight organism turn brown rapidly and look burned. Positive identification of the two diseases cannot be made without laboratory tests. The necrotic leaf lesions made by the bacterial wilt organism are hard to tell from those caused by the common and fuscous blight organisms. From a combination of all the symptoms of bacterial wilt, this disease can usually be told from the other three diseases.

Water-soaked spots similar to those on the leaves are produced on the pods by all of the organisms, except the wilt organism. The spots grow larger, and the tissue around each is reddish brown or brick red. The spots finally dry and are often covered by a dried bacterial ooze. Commonly the bacteria invade the upper suture of the pod, cause a water soaking of the tissue on both sides of it, and later infect the seed. On light-colored bean seed a yellowish discoloration may be visible at the hilum, the point of attachment to the pod, but on colored seed the infection is hard to detect. Badly infected seeds usually are shriveled and yellow. Slightly infected seeds may appear healthy.

Upper left: Stem anthracnose on lima bean pods. Upper right: Downy mildew on pods of lima bean. This organism requires a high humidity to cause infection. Lower left: Bacterial blight infection of pea leaves. The bacteria are spread from plant to plant by rain and hail. Lower right: Red node, a virus disease of bean. The only evidence of infection of pods by the bacterial wilt organism is a slight discoloration of the upper suture, but the seeds inside them often are yellow and shrunken. When the infected seeds are planted, the plant becomes infected internally, and the bacteria spread through it. General stunting and death may result. Some of the seedlings may grow to fairly good sise, but usually a lesion, known as stem girdle or joint rot, occurs at the node where the cotyledons were attached. The affected stem, further strained by the increasing weight of the top, breaks at the node.

The bacteria that cause common blight (Xanthomonas phaseoli), fuscous blight (X. phaseoli var. fuscans), and halo blight (Pseudomonas phaseolicola) enter the plants through the stomata of the leaves, stems, and pods. The wilt organism (Corynebacterium flaccumfaciens) enters through wounds. Rain, hail, driving winds, damp weather, and even sprinkler irrigation favour the spread of the organisms.

Plants grown from infected seed frequently develop lesions on the cotyledons and stems. From those spots bacteria are splashed by rain or hail to leaves of other plants. A few diseased plants scattered through a field as a result of infected seed may be the source of a general outbreak if weather conditions are favourable for the spread of the organisms. Warm weather favours common and fuscous blights and bacterial wilt. Cool weather favours halo blight.

The bacterial blight organisms can also live over winter on diseased bean refuse in the soil, possibly for 2 years. The most effective control measure against bacterial blights is the use of disease-free seed. One should use only seed grown in areas where rainfall is scant and blight epidemics do not occur.

Because the blight organisms can overwinter in the soil, beans should not be planted for at least 2 years in fields that have been infected. Nor should bean straw from a crop that showed blight be applied to land that is to be planted to beans. Infected bean straw used for bedding farm animals may continue to harbor bacteria, and the use of such straw on fields to be planted to beans is unwise.

Bacteria adhere to clothing and farm machinery. Picking, cultivating, and similar operations can spread the bacteria from a few infected plants to the rest of the plants in a field. No variety of snap beans or dry beans is resistant to common or fuscous blight. Many of the varieties of dry beans, such as Pinto, Great Northern, Red Mexican, and Michelite, are highly resistant to halo blight, but all snap beans are susceptible to it. Not much is known about resistance of any of the varieties to bacterial wilt. Seed treatment is of doubtful value in the control of the blights. Spraying or dusting the plants with fungicides or bactericides has not proved to be a practical control measure. Rust is one of the most serious diseases of dry beans in irrigated sections of

Colorado, Nebraska, Wyoming, and Montana some years. It also causes losses in pole snap beans along the coast of southern California, Oregon, Washington, and Florida, and in other States. In 1950 and in 1951 it caused considerable damage in western Nebraska. It was widespread in 1951 in northeastern Colorado, Bean rust infects only beans. None of the rusts of the other crops infects beans.

The bean rust organism commonly attacks the leaves. It attacks the stems and pods less commonly. The first symptoms are small white spots or flecks on the under sides of the leaves. In a few days the spots break open into rusty-colored lesions, which occur on both leaf surfaces. The leaf begins to yellow about a week after the lesions appear. Later it turns brown and dries up. In a seriously infested bean field the plants look as if they had been scorched.

A fungus, Uromyces phaseoli typica, causes bean rust. It reproduces by spores, of which there are several types. The organism cannot be grown on any known artificial media as many other fungi can be grown. It grows only on bean plants. Usually the grower notices only one of the spore stages, the summer stage. It shows up as reddish-brown pustules, which contain thousands of spores. The spores, called uredospores, readily become detached and may be blown by the wind from plant to plant and field to field. Towards fall the rust produces another type of spore, which is black and known as a teliospore. These spores live over winter, mainly on old bean straw. A small per centage of the summer spores also survive the winter.

The teliospores germinate in spring. Each produces four smaller spores of still another kind, called sporidia. They are blown about by the wind. If conditions are favourable when they alight on a bean plant, they germinate and grow into the leaf. Later they produce a fourth kind of spore, known as aeciospores. These, in turn, infect bean leaves and produce the red summer spores.

The aeciospore stage is found only in some bean-growing areas, particularly in the far West we do not know why. It has never been reported in the intermountain region. But there the small per centage of uredospores that overwinter may also be blown by the wind and later germinate and penetrate into the leaf of a young bean plant. Some days later occasional white flecks may appear on the under side of the leaves. The flecks soon break through the leaf surfaces and produce the rusty pustules of the type the grower observed the previous summer.

Since each pustule produces thousands of spores, the disease may develop readily if weather conditions are right. Each spore in turn may produce another pustule on a leaf in about 10 days. The cycle may be repeated for 5 to 6 weeks. Unless steps are taken to control the disease an epidemic may occur and the crop may be lost. Bean rust develops most rapidly when the humidity is high, as the result of rain, dew, or irrigation. Heavy vine growth, which shades the ground and prevents air circulation, produces ideal conditions for rust. Bean rust is not seed-borne.

The spores live over winter on old bean straw and the disease may get started in the field if beans are planted on land that produced a crop of rust-infected beans the preceding year or on new land on which infected straw was spread. The spores on the straw may germinate and infect the new crop. The disease might also get started if beans are grown close to stacks of old, infected bean straw. Spores may be blown to nearby bean fields and infect the plants. More than 30 races or strains of bean rust have been identified. Each reacts differently on different varieties of beans.

Some varieties are susceptible to certain races but are resistant to others. No variety that resists all the known strains of rust has been developed. The few noncommercial varieties that resist a great many of them are used as parents in the breeding of new varieties resistant to as many strains as possible. The most satisfactory artificial control is a dust of finely ground sulfur. It can be applied to the plants with a power duster on the ground or from an airplane. The dusting should be done fairly early in the season before the rust spots become numerous.

A prompt application destroys the few pustules present at that time, stops the spread of spores from the spots, and prevents the production of others. If dusting is done after the rust is widespread, more applications are required and the control is not nearly so satisfactory. If the disease is widespread in a district, two or three dustings at intervals of 7 or 10 days are necessary. The best time to dust is when the wind is not blowing and the plants are dry. About 25 to 30 pounds of sulfur to an acre should be used for each dusting.

Bean rust can be almost eliminated in a district in a few years if growers cooperate and dust their beans with sulfur. This must be started fairly early in the season and before the plants overlap in the rows usually about 6 weeks after planting. All commercial varieties of dry beans are susceptible to most of the races of rust. Men in the Department of Agriculture developed two Pinto varieties, No. 5 and No. 14, which resist many of the races but are slightly too viny for commercial use. Other resistant types of Pinto, of more of a bush habit, and several Great Northern types should be ready for distribution to growers in a few years. Most of the bush snap beans are highly tolerant of rust. Some of the pole snap beans, however, such as Blue Lake, McCaslan, and Kentucky Wonder are very susceptible. Among the tolerant pole beans are White Kentucky Wonder, the U. S. No. 4 Kentucky Wonder, Potomac, and the Rialto.

Root Rots occur wherever beans are grown, but they cause more damage in the Southern States than elsewhere. They have become a problem in parts of the West. The organisms causing root rots live in the soil and attack beans when soil and weather conditions are unfavourable for best plant growth.

The organisms as a group form lesions or cankers on the stem below the soil level, on the taproot, and on the fibrous roots. The cankers, which are of various sises and shapes, may may be gray, black, brown, or brick red. Decay of the fibrous roots is common. Infected plants generally are stunted and somewhat yellowed. The dry root rot organism usually causes a reddening of the taproot, which may extend to the soil line. The rootlets that develop from the taproot are killed. The main root and the lower part of the stem beneath the soil are often pithy and dry.

The rhizoctonia root rot organism causes reddish-brown, sunken cankers on the roots and the stem below the soil line. The cankers often girdle the stem. The pythium wilt and southern wilt organisms cause sudden wilting and death of the plant. Pythium wilt usually occurs on the stem at the soil line. The affected tissues become soft and slimy. The infection often extends upward into the branches, but it does not extend much below the soil surface.

Southern wilt rots roots and stems, which later become dry and papery. A white fungus growth develops on the stem at the soil surface and grows into the soil around the plant. When an infected plant is pulled from the soil, a collar of fungus mycelium and soil often clings to it. The causal organisms of dry root rot (Fusarium solani f. phaseoli), rhizoctonia root rot (Rhizoctonia solani), pythium wilt (Pythium buderi), and southern wilt (Sclerotium rolfsii) can live for fairly long periods in the soil where no cultivated crop is being grown. All the organisms, except the dry root rot organism, cause root rots of other crops. The southern wilt organism is found primarily in the South, but the other organisms normally are present in nearly all soils. They like cool weather and high soil moisture.

None is seed-borne. We know of no effective artificial control measures for the root rots. No varieties are resistant to any of them. Cultural practices that improve the general growing conditions of the crop will reduce the severity of the diseases. The most effective control is a 4- to 5-year rotation that includes cereals, clover, and alfalfa plants on which the root-rotting organisms rarely cause damage. Anything that makes conditions favourable for bean plants tends to make conditions unfavourable for the fungi. Sclerotinia wilt, or white mold, is a serious disease of beans in several parts of the United States in parts of southern Florida, Wyoming, Montana, Idaho, and Oregon.It affects the stems, leaves, and pods.

Small, soft, watery spots appear first. The organism grows rapidly. If several days of cool, wet weather follow the infection, a cottony fungus growth spreads over the branches and leaves. Heavy vine growth in irrigated fields causes the plants to stay moist for a rather long time, and so the fungus becomes active. The affected tissues dry out and become light, punky, and often severely shredded. Irregular, hard, black bodies, one-fourth to one-half inch in diametre and called sclerotia, frequently occur about that time on infected tissue. Those are the resting bodies of the fungus. Severely affected plants usually die within a few days.

The fungus Sclerohnia sclerotiorum causes sclerotinia wilt. The black sclerotia fall to the ground; when moisture and temperature are right, strands of white mold grow out of the bodies and infect bean plants directly. If conditions are not favourable when the sclerotia fall to the ground, they may remain in a resting stage for as long as 10 years. Infection takes place also by spores. From the sclerotia grow tiny bodies, like mushrooms, each of which may produce thousands of spores. Spores that are blown to nearby bean plants germinate and infect them.

Soon new masses of mold appear. In them more sclerotia are formed. As each sclerotium can produce many spore-bearing bodies and each of the bodies may produce thousands of spores, even a few sclerotia in one field can sometimes cause widespread infection if conditions are suitable. No bean varieties are known to be resistant. No spray or dust materials have been found to control the disease satisfactorily.

A high per centage of the sclerotia can be destroyed by flooding infected fields for 3 weeks or more. In parts of Florida where that practice is feasible, the best season for flooding is summer because high temperatures hasten decay. Infested bean straw and cull beans contaminated with sclerotia should not be fed to animals if the manure is to be used on land where beans may soon be planted. When possible, infested bean fields should be planted for 2 years or more to crops including small grains, corn, and hay that are not susceptible to sclerotinia wilt. That practice is not likely to eliminate the trouble entirely, but it will reduce losses. Common bean mosaic has been known for more than 50 years and is world-wide. Since 1930 several other mosaics of bean have been reported. Because the different mosaics generally produce similar field symptoms, they are discussed here more or less as one disease.

Many growers underestimate the importance of bean mosaics because infected plants are rarely killed. The mosaics reduce yield and quality of the product. The annual losses from these diseases total several million dollars in the United States. The important mosaic diseases of bean are common bean mosaic, a variant of that disease called New York 15 mosaic, and yellow bean mosaic. Two other bean mosaics, southern bean mosaic and pod mottle, have been found, and may become of economic importance if they become widespread. Several strains of yellow bean mosaic have been identified. In some years they have caused considerable damage.

Common bean mosaic is the most widespread of the bean mosaics. The New York 15 mosaic virus, which produces much the same symptoms, is less serious. Yellow bean mosaic, which is primarily a disease of sweet-clover and is found also on crimson clover, red clover, and gladiolus, has entirely different symptoms. Plant lice spread it from those crops to beans. The bean mosaics cause mottling and malformation of leaves and stunting of the plants. The symptoms are more intense at high temperatures.

Symptoms are not noticed on stems or seeds. The patterns made by the mottled yellow and green areas may vary. Dark-green areas frequently develop near the veins and veinlets of infected leaves. Leaves of very susceptible varieties may be considerably puckered and may develop areas that look like blisters And warts. On less susceptible varieties, the leaves are ruffled, crinkled, or cupped downward. With yellow bean mosaic, the contrast between the yellow and green areas of the leaves usually is more intense than it is with common mosaic. On solve resistant varieties the common mosaic virus may produce occasionally a necrosis of the leaves, pods, stems, and roots. This symptom has been observed only rarely in the field.

Pods from infected plants are sometimes mottled, deformed, and rough. Some may be very shiny. Yellow mosaic (especially the pod-distorting strain), southern bean mosaic, and pod mottle viruses produce the worst symptoms on pods. Each of the bean mosaics is caused by a different virus. Plant lice usually spread the viruses from diseased to healthy plants. The viruses of the important mosaic diseases are not spread from infected to healthy plants by cultivation or picking, but the viruses of southern bean mosaic and pod mottle can be spread thus. The virus of common bean mosaic and its variant are seed-borne.

The infected seeds look normal, but they carry the virus and produce diseased plants. None of the viruses causing the other mosaic diseases is carried in the seed. The only satisfactory way to prevent mosaic is to use resistant varieties. None resists all viruses. Some of the bush snap beans that resist common bean mosaic and its variant are Contender, Florida Belle, Logan, Improved New Stringless, Idagreen, Puregold, Rival, Idaho Refugee, Sensation Refugee 1066, Sensation Refugee 1071, U. S. 5 Refugee, Tender-long 15, Ranger, Toperop, and Wade. The resistant pole varieties include Blue Lake, Kentucky Wonder, and U. S. No. 4 Kentucky Wonder.

Dry beans resistant to common bean mosaic are Michelite, Robust, Great Northern U. I. Nos. 16, 31, 81, and 123, Montana Nos. 1 and 5, Red Mexican U. I. Nos. 3 and 34, Pinto U. I. Nos. 72, 78, and 111. Only the Great Northern strains are resistant to the New York 15 variant strain. No variety has yet been developed that resists yellow bean mosaic. In sections where yellow bean mosaic is widespread it is a good practice to avoid planting beans too close to fields of sweetclover, red clover crimson clover, and gladiolus. Fence rows and ditch banks should be kept free of sweetclover, which commonly grows "wild" in many Western States.

Curly top is a virus disease of beans and several other crops. It commonly is found in Utah, Idaho, Washington, Oregon, and California, and sometimes in other Western States. In some localities it is not serious every year, although when there are many beet leafhoppers, which transmit the virus, the bean crop may be ruined. In some parts of Idaho, Washington, and Oregon, the disease has been so serious every year that only resistant varieties survive.

Young plants infected with curly top are decidedly dwarfed. The symptoms are most pronounced on the trifoliate leaves which pucker, curl downward, and show a clearing of the veins. The young leaves cease to develop, turn yellow, and curl downward. They are thick and brittle and readily break off from the stem. Frequently they are cupped; sometimes they resemble small green balls. Plants infected when very young usually die; plants that are infected later do not die, and typical symptoms do not always develop. The pods produced on infected plants are usually stunted. The disease is not seed-borne.

Curly top is spread from plant to plant only by the beet leafhopper, usually called whitefly in the West. Before being able to infect beans, it must feed on some plant that is infected with curly top. It overwinters in desert areas where many weeds carry the curly top virus. In the spring the leafhoppers migrate to cultivated fields of beans, beets, tomatoes, and other crops and spread the disease. The symptoms appear about 10 to 14 days after infection. Within a few weeks after that an infected plant may die. Because curly top is transmitted only by leafhoppers, the disease is prevalent only where that insect thrives. The severity of the disease reflects the population of the leafhoppers and the per centage carrying the virus.

Curly top can be best controlled by growing resistant varieties. Unfortunately, no commercially suitable resistant variety of snap bean has been developed, but progress is being made in obtaining such beans, notably in developing several resistant Blue Lake types. Resistant varieties of dry beans are Great Northern U. 1. Nos. 16 and 31, Red Mexican U. I. Nos. 3 and 34, and Pinto U. 1. Nos. 72, 78, and III, Pink, and Pioneer. Pioneer can also be used as a snap bean in home gardens in regions where curly top is prevalent, although as such it is only fair.

Downy mildew was first reported in the United States about 60 years ago. In later years it has caused considerable damage to lima beans in some of the Middle Atlantic and North Atlantic States. It is of minor importance elsewhere. The fungus that causes it develops best during periods of wet weather with cool nights, heavy dews, and fairly warm days.

The disease is recognised by the white, cottony growth that forms in large, irregular patches on the pods. Each patch may have a purplish border. When young pods are attacked and when older ones are covered with fungus growth they shrivel, wilt, and die. Downy mildew less frequently attacks young leaves, shoots, and flower parts. Infected leaves may show irregular spots, particularly near the veins, without much visible fungus growth.

Downy mildew is caused by a fungus, Phytophthora phaseoli, which attacks only lima beans. The organism may penetrate the seeds and live in them over winter. It can also survive in debris which may be a source of infection of the new crop. During the growing season, the fungus may be spread to healthy plants by insects that visit infected flowers. Rapid natural spread depends on ideal weather conditions favouring the fungus.

Seed from an infected crop should not be used for planting. Seed produced in the far West where downy mildew is not found should be used. A 2- or 3-year crop rotation is wise. When the disease is first noted the crop should be dusted with a copper dust containing 5 to 7 per cent actual copper, such as tribasic copper, at the rate of about 40 pounds of dust to the acre for each application. Dustings should be made at weekly intervals as long as the weather favours the development of the disease. No commercial variety of lima beans is resistant. Several resistant noncommercial types have been found. A breeding programme has been started to incorporate that resistance into suitable varieties.

ROOT ROTS, WILTS, AND BLIGHTS OF PEAS

The pea (Pisum sativum) is subject to various types of diseases blights, root rots, and wilts. The formulation of control measures for them requires knowledge of their symptoms, causes, relations of host and parasite, and possible inherent resistance. Blight refers to the discoloration, gradual drying, and eventual death of the affected plant parts. Bacteria and fungi cause blights, which usually are seed-borne and develop best during wet weather.

Only one major bacterial blight (Pseudomonas pisi) affects peas. A number of fungal blights, however, do occur on peas, among which are septoria blotch (Septoria pisi), downy mildew (Peronospora pisi), powdery mildew (Erysiphe polygoni), anthracnose (Colletotrichum pisi), ascochyta blight, and several minor ones. Ascochyta blight is among the oldest and worst. Its nature and control are much like those of other fungal blights (except powdery mildew). Details about it therefore apply generally to the others.

Ascochyta blight is a composite of three diseases resulting from infection, singly or collectively, by Ascochyta pisi, A. pinodella, or Mycosphaerella pinodes. The nature of the disease was first determined in Europe in 1830. It was a constant threat in the United States until about 1915, when it declined with the shifting of the seed industry to areas in the West where the low rainfall before and during the harvest reduced the amount of infected seed. But in some years unseasonal rainfall during harvesttime in western seed-growing districts means contaminated seed, so that the disease remains something of a problem.

The symptoms the three parasites cause are almost alike. Most evident are the lesions that begin as small purplish specks on leaves and pods. When infection is caused by M. pinodes or A. pinodella, the specks may enlarge on the leaves into round, target-like spots. If numerous, they join to make irregular brownish-purple blotches on both leaves and pods. A. pisi produces relatively few, rather definite, sunken, tan or brown spots, which have dark-brown margins and are circular on leaves and pods. The pod lesions usually become sunken.

Stem lesions are elongated and purplish black. They originate as separate infections or as a continuation of petiole infection around the nodal area. They also may coalesce and girdle the entire stem, weakening it so that it is easily broken. Affected leaves eventually shrivel and dry into a blighted condition, which resembles freshly cured clover hay. M. pinodes may also blight the blossoms and young pods and cause withering, distortion, and eventual dropping. All three organisms can attack that part of the stem and root at the soil line and produce a bluish-black foot rot. It is severest when it is caused by A. pinodella. A. pisi seldom causes severe foot rot.

The three organisms responsible for the ascochyta blight complex are closely related, but each has marks that classify it as a distinct species. In pure culture on artificial media, the light-colored mycelium and the abundant carrot-red spores of A. pisi readily distinguish it from the darker-colored mycelium and relatively scarce light-buff spore exudate of M. pinodes and A. pinodella. All produce waterborne spores, the pycniospores, but those of A. pinodella are only half as large as the spores of the other two species. The incubation period of the disease caused by M. pinodes and A. pinodella is 2 to 4 days, compared with 6 to 8 days for A. pisi. M. pinodes is the only species that produces ascospores. Such spores can be carried for considerable distances by air currents and largely account for the more aggressive nature of that pathogen.

The three disease organisms can infect the seed. Such infection serves to overwinter the pathogens and is a means of transporting the disease from one region to another. The pathogens also overwinter in infected pea straw. In regions of extremely mild winters, they may remain active on infected volunteer plants. The ascochyta blight organisms do not live indefinitely in the soil as in the case of the root rot and wilt fungi, but remain there only as long as the infected pea straw is not completely decomposed.

When infected seed is planted, the disease first appears as a foot rot on the young seedlings at the point of seed attachment and often kills or weakens the young plants. Spores are produced during wet weather on such plants and spread the disease to nearby plants. Infected pea straw in the soil from the crop of the previous season may give rise to both pycniospores and ascospores. As pycniospores need spattering rain for their dissemination, they spread the disease only a few feet from the source.

Ascospores, however, are shot out from the spore-bearing structures in the old plant tissue and are carried quite far by air currents. They usually are more abundant than pycniospores and spread the disease uniformly over the field rather than in small patches. If conditions are favourable, both types of spores may be formed continually on the dead parts of the infected plant. Because the ascospores are more widely and uniformly distributed, however, infection by M. pinodes (the only species producing ascospores) is more damaging. Because moisture is required for spore discharge and infection, rainfall, dews, and high humidity are the most important environmental factors in the development of ascochyta blight. The number of periods of wet weather during the pea season largely determines how bad the disease will be.

Because there is no practical resistance in the pea to the ascochyta blight complex and because nothing can be done at present to change the weather, the disease is best controlled by avoiding, eliminating, or reducing the causal organisms by the use of disease-free seed, crop rotation, and sanitation. It is unwise to plant seed grown in humid sections of the East and Middle West. That seed is likely to be infected with the disease organisms.

Furthermore, the straw from a seed crop is apt to be more heavily infected with the organisms than a crop cut in the green stage for processing. That increases the inoculum potential on the overwintered straw and affords a better chance for the disease to establish itself earlier and more severely. Seed grown in drier regions of the West is the best insurance against seedling infection. Treatment of seed with fungicides reduces the surface contamination but will not eliminate internal infection.

Rotation to control pea blights implies more than merely avoiding planting pea crops on the same land more than once every 4 or 5 years. Such a rotation undoubtedly would eliminate the organism from the soil. But, in addition, every effort should be made to locate new plantings as far as possible from those of the previous season. Even though most of the vines from the old fields are removed to the viner station, there may be enough infection on the stubble and debris to provide for the dispersal of ascospores the following spring.

Of equal importance is sanitation. Any infected plant parts whether Stubble in the field or vines on the viner stacks offer a constant source of inoculum for the next crop. After the ensilage has been removed from the stacks, the outside part of the stack, which usually is raked off, should be destroyed before seeding the next spring. If mobile viners are used in pea fields, the vines and pods, instead of just the stubble, are left in the fields.

That will create a situation comparable to one in which seed peas and processing peas are grown in the same area, for the fresh vines will dry, and the organisms will continue to develop, thereby increasing the inoculum potential for the next season. A good practice would be to plow down all pea stubble and vines immediately and plant the field to a crop, such as grain, in which the soil will not be cultivated during the next season. In some places it is customary to plant éarly varieties alongside late ones or to stagger planting dates. In either event, ascospores may develop on the stubble of the earlier planting and provide for abundant dispersal of spores to the younger plantings nearby. Such practices should be discouraged.

Root rot is caused by a number of fungi, which singly or together attack the cortex, the tissue outside the water-and food-conducting cylinder of the roots and lower stem. Invasion of the cortex may be quite general, or it may be somewhat localised. The rot may be soft and watery, or it may be more like a dry, corrosive decay, depending on the causal organism. The various types of root rot are designated by the names of the fungi that cause them, such as aphanomyces root rot, fusarium root rot, and ascochyta root rot.

The damage done by root rot varies With the season and the causal organism. One year the crop may fail; the next year on the same piece of land the crop may be all right. The disease may act as a seedling blight and kill scattered plants at an early stage, or it may not attack the plant until quite late in its development. Two different pathogens attacking the same plant frequently cause more harm than either alone. Complete crop failures can occur. More often, however, yields are reduced in varying degrees as a result of an impaired or restricted root system. Under those conditions, the affected plants look as though they were suffering from malnutrition.

The disease pattern or symptoms and the conditions favourable for root rot development depend upon the specific pathogen. Practically all the types may occur wherever peas are grown, but certain ones are more prevalent in some areas than in others. Aphanomyces root rot, caused by Aphanomyces euteiches, occurs to some extent wherever peas are grown and is often referred to as common root rot. It is especially prevalent in the older pea-growing regions of New York, New Jersey, Wisconsin, and Minnesota.

Some 10,000 acres of canning peas were lost in 1942 in Wisconsin as a result of the disease. The disease is first detected by a soft and watery condition of the stem an inch or two above the soil line. By then the roots have become similarly affected. The diseased tissue becomes discolored because other fungi invade the softened tissue. Ordinarily most of those fungi cannot infect the healthy tissue. In time the affected tissue on the stem above the soil line collapses and shrivels. The outer tissues of the roots then are so rotted that when the plant is pulled all that remains of the root system is the stringy central core of the taproot. That condition usually distinguishes the aphanomyces from other root rots in the field. Severe infection of young plants usually kills them before they blossom. Plants infected later seldom die but, because of a restricted root system, they are stunted; their leaves turn yellow and die from the ground up. The fungus that causes aphanomyces root rot belongs to a group of fungi commonly called water molds. It has two kinds of spores. One, the oospore, or resting spore, is the thick-walled spore generally found in diseased tissue. It may also occur in culture on artificial media. The other, the zoospore, or free-swimming spore, has two long hairs, which enable it to move about in water.

Very likely the nutrient concentration in the soil water affects hyphal growth and spore germination similar to that observed on artificial culture media. Aphanomyces root rot can occur on relatively dry soils, but it never becomes severe unless the soil is wet either as a result of a high water-holding capacity, as in clay soils, or frequent heavy rains. Severity of infection is associated generally with a large number of infection points on the root system. In relatively dry soil the nutrients are concentrated in the soil water; then few zoospores are apt to be produced and the infection points may be scarce. In wet soil, however, the soluble nutrients are leached or diluted and abundant spores may form, so that severe disease develops. That supposition is strengthened by the fact that commercial fertilisers, especially those high in nitrogen, sometimes aid in controlling root rot. The nitrogen fraction contributes heavily to the soluble salt content of the soil.

Fusarium Root Rot is serious in Colorado and the Pacific Northwest. In some places, New York among them, it and aphanomyces often occur together, and then the damage is worse than that caused by either organism alone. The disease usually starts on the seedling at the junction of the root and stem where the germinated seed remains attached. From there it extends upward as a wedge-shaped, reddish-brown lesion. The decay also may extend downward and involve primary and secondary roots. There is no pronounced water soaking of the invaded tissue, and the discoloration is immediately apparent.

Further progress of the disease causes the tissue to shrink. In time the stem is girdled, and often the plant breaks near the place of seed attachment. A reddish discoloration of the central water-conducting vessels may occur, but usually only in the zone of seed attachment and seldom more than an inch beyond. If the main root is severed from the plant in the seedling stage, the plant attempts to maintain itself by sending out new roots through the underground part of the stem, but eventually it dies or is severely stunted. Many species of Fusarium have been described as causing a root rot of pea, but the generally accepted causal organism is F. solani f. pisi. Fusarium root rot occurs in fairly dry soils as well as wet soils. It flourishes when the soil temperature is around 80 F. It develops most severely when the spring planting time is warm or when the peas are planted late.

Rhizoctonia root rot of pea is primarily a disease of the seedling stage. It usually is of minor importance. The causal fungus is the sterile or Rhizoctonia stage of Pellicularia filamentosa. It causes injury to a wide variety of farm crops, particularly young seedlings and potato sprouts. The same fungus causes black scurf of potato tubers.

On peas, in the early stages, the disease appears first as a slightly eroded, yellowish-brown zone on the underground stem and root in the region. of the seed attachment. The tissue later turns a darker brown, and the lesions become sunken and heavily eroded. They may girdle the stem. More frequently, however, it attacks the growing tip of the emerging seedling, killing it before complete emergence. When that happens a new shoot is sent up, which also may be killed, so that the affected seedling may have a number of growing shoots with dead terminals. That is quite characteristic of this root rot.

Temperature is the primary factor in the development of this disease on pea. A soil temperature of 60 to 65 F. is most favourable. Various species of Pythium, notably P. ultimum, can parasitise the tissues of the underground part of the pea plant. In the field, however, most of the damage from the organism results from seedling injury or seed decay. Affected plants show a water-soaked, somewhat translucent and softened tissue extending above and below the region of seed attachment. As the disease progresses, the tissue becomes more and more discolored. It develops best in wet soil.

It is unlike Aphanomyces in being most aggressive during the seedling stage it rots the germinating seed or kills the emerged seedling. If the plant survives and conditions are not unusual, Pythium um by itself has little effect on root rot. High temperature, accompanied by high humidity or wet soil for a long time, may incite the organisms to further decay of the root system. Control of the root rot complex of pea is difficult because of the nature of the different causal fungi, most of which live indefinitely in the soil. No resistant varieties are available. In infested areas, growers have to depend on seed treatment, rotation, and good management of soil and crops.

Seed treatment offers protection against some phases of some of the root rots. It largely controls the seed decay and damping-off stages that result from infection by Fusarium, Pythium, and Rhizoctonia, but it does not preclude infection beyond those stages. Aphanomyces root rot, however, is not affected by seed treatment. The main purpose of seed treatment is to provide for healthier, more vigorous plants which are better able to ward off the effects of later infection. Crop rotation, properly carried out, keeps the pathogens from accumulating in the soil. Because some of the pathogens; including Pythium, Rhizoctonia, Aphanomyces attack different crop plants, rotation may seem of little value against them.

Actually, though, a nonsusceptible crop immediately preceding the peas is of some value. Surveys of pea fields affected by Aphanomyces and Fusarium have demonstrated that the severity of root rot declines as the interval between pea crops increases.

The main purpose of rotation, however, should be to prevent the build-up of the fungi in relatively healthy soil by 4- to 5-year intervals between crops rather than to attempt to reduce the pathogens after the soil has become heavily infested with them. Good management of soil and crops deters the establishment of the fungi and enables the crop to produce if disease is present. High soil moisture favours aphanomyces root rot and others to some extent. Soils with good external and internal drainage should be selected therefore for peas. One should not return diseased pea straw to the soil immediately. Because peas affected by root rot have a restricted root system and behave as plants suffering from malnutrition, one should maintain high fertility in the soil.

Applications of fertilisers sometimes make a crop possible despite aphanomyces root rot. In New York applications of commercial fertilisers at the rate of 50 to 100 pounds of nitrogen to the acre have given good results. At times, especially in bad disease years, the increase in yields does not justify the added cost of fertilisers. Gypsum at rates of one-half ton to 1 ton an acre have reduced aphanomyces root rot but not fusarium.

Pea Wilt is one of a group of diseases that are incited by certain vascular Fusaria. It used to be confused with the root rot complex generally attributed to "pea-sick" soils. It differs from the root rots in that the cortical tissue is seldom affected under field conditions. Rather, the fungus establishes itself in the central core of water-conducting vessels and produces a toxin that causes a progressive yellowing and wilting of the foliage. Death of the plant may ensue.

The two distinct fusarial wilts of pea are wilt and near-wilt. Fusarium wilt of pea was first discovered in Wisconsin in 1924. Since then it has been found in the major pea areas of the United States. Its first noticeable symptom is a downward curving of the stipules and leaflets and a slight yellowing of the leaves. A superficial grayness like waxy bloom may also occur. The lower internodes thicken and the entire stem becomes somewhat rigid. Thereupon the plants may wilt abruptly at the top, followed by a shriveling of the stem. If soil temperature is low, such wilting may not occur. Instead the affected plant becomes yellow and withers slowly, leaf by leaf, from the ground upward. The central water core turns yellow or orange.

Symptoms often occur in advance of the fungus in the vessels. At times wilt symptoms appear with very little vascular discoloration. The cortex, the outer part of the root and stem, is usually quite sound, unless root rot fungi enter it. In a newly infested field only a few scattered plants may show infection the first year. With repeated plantings to susceptible varieties the infestations appear as enlarged round areas, which eventually merge until the entire field becomes infested. The disease can spread rapidly.

Fusarium wilt of pea, caused by Fusarium oxysporum f. pisi race 1, exhibits features not altogether common among diseases caused by other forms of the species. But like the others the pathogen inhabits the soil and, once established, remains there indefinitely. The best temperature for growth of the fungus on artificial media is about the same as for other vascular Fusaria; about 80 to 85 F. The best temperature for disease development in soil differs from that of the other Fusaria.

With most of them, disease development follows closely the optimum temperature for growth of the fungus, but in pea wilt it is distinctly lower, about 70 . If, however, peas are grown in nutrient sand culture and the sand is artificially infested with the pathogen, the optimum temperature for disease development is not different from that for the growth of the fungus. This suggests that a biological antagonism occurs at the higher soil temperature, which interferes with the infectivity of the pea wilt pathogen. That possibility is supported by earlier demonstrations that fusarium wilt would not establish itself in certain types of Wisconsin soils.

Fortunately for the pea industry in districts where the disease appeared, a heritable resistance was found among some varieties in the field. The resistance, a clear-cut character governed by a single dominant gene, was soon incorporated into desirable canning and market varieties. Today virtually every canning pea variety possesses wilt resistance and the disease is completely controlled. After wilt-resistant varieties had been established, another vascular wilt appeared. It was discovered in Wisconsin in 1931 and was called near-wilt, because it resembled the old wilt in certain respects.

Soon the same or similar diseases were reported in other regions. The near-wilt pathogen, F. oxysporum f. pisi race 2, has much in common with the wilt pathogen. It remains indefinitely in the soil. Its best growing temperature is the same as for the wilt fungus. It is unlike the wilt pathogen in that the best temperature for disease development in the soil is higher and follows that for the growth of the fungus.

Near-wilt is worse in warmer seasons, on late varieties, and on late plantings. The fungus that causes near-wilt establishes itself on all soil types. It enters the water-conducting vessels of the pea through either the root tips or at the cotyledonary node. The wilt fungus does not progress far enough up the stem to infect the seed, but the near-wilt fungus frequently travels the entire length of the stem. As a result, seeds (especially those of the dwarf, late varieties) are likely to become infected. The fungus can be disseminated to other fields by infected seed and surface-contaminated seed and by spreading diseased vines on the soil.

Workers at the Wisconsin Agricultural Experiment Station discovered one line of breeding material that appeared to have complete field resistance to near-wilt. The resistance is inherited as a single dominant gene. Delwich Commando was introduced as the first variety completely field-resistant to wilt and near-wilt. Bacterial blight, although caused by a bacterium, behaves like some of the fungal blights. The disease occurs on all above-ground parts of the plant, but is most pronounced on the leaves, stems, and pods.

The symptoms first appear as small, irregular water-soaked spots, which may enlarge to one-eighth inch across on the foliage and to about one-fourth inch on the pods. Streaks usually appear on the stems. The translucent leaf lesions usually turn to a golden-brown colour. In time the lesions coalesce. The leaves wither and become papery. The stem lesions elongate and remain water-soaked or turn brown with age. On the pods the spots become slightly sunken but retain the water-soaked character. They may occur on the checks of the pods or along either suture. The causal organism, Pseudomonas pisi, is carried over winter chiefly through the seed, either as a contaminant on the seed surface or internally as an infection. The organism can infect cowpea, sweet pea, hyacinth bean, and the perennial or everlasting pea. Possibly the bacterium overwinters in the vines of those hosts, although the seed-borne character is most important.

Spread and damage depend on weather. Spattering rains carry the bacterium from one plant to the next. Seedling infection may kill the plant. Usually, however, such infection provides the inoculum for spread to adjacent healthy plants. The extent of spread depends upon the frequency of rainy periods. If' they occur often enough, within the span of a week, the primary infection from a tiny amount of infected seed can spread over a large field. In continued rainy periods the crop might be destroyed. Dry weather may check the disease. The bacterium enters the plant through the stomata and through wounds. Any practice, therefore, that injures the plants, especially when they are wet, serves to increase infection. Hail injury frequently paves the way for rapid and severe infection. The most effective control of bacterial blight is the use of disease-free seed. A 4 or 5-year rotation, recommended for the control of other pea diseases, would certainly eliminate diseased vines as a primary source of infection.

BLIGHTS AND OTHER ILLS OF CELERY

Celery is grown extensively as a truck crop on the muck lands and irrigated mineral soils of many States from Florida to Massachusetts, the Great Lakes States, and some of the Mountain and West Coast States. It is grown also as a market garden crop near many large population centres. It has an annual value of more than 50 million dollars. In some districts celery has been grown intensively for nearly a century.

The importation of much of our seed from Europe before 1920 and the free exchange of seeds and plants within the States at all times have meant that there are few if any celery diseases that we have not acquired. Appropriate control measures have been developed for many of them by Federal and State agencies and seed growers. The most widely distributed and costly diseases of celery are the early and late blights, caused by the fungi Cercospora apii and Septoria apii var. graveolentis, and the rather minor bacterial leaf spot, caused by Pseudomonas apii.

Cercospora may be seed-borne, hence early blight occurs almost everywhere that celery can be grown. Because it requires hot weather for its most rapid development, the fungus is most troublesome on the early summer crops in the Northeastern and Great Lakes States. In Florida it may occur in the seedbeds in October and do some damage all winter, but the greatest losses occur on the late winter crop, which matures during the warm weather of March to May. It is checked by periods of cool weather below 40 F.

Losses are due to stunting of growth, the necessity for heavy stripping of diseased stalks at harvest, and poorer keeping and market quality. The fungus overwinters readily on debris from a previous crop. Early blight first appears on seedlings in the plant bed or on transplants in the field as small, pale-green or yellow spots a week after inoculation with the spores of Cercospora. The spots enlarge and often envelop much of the leaf. They turn brown to slate gray as the fungus fructifies on the lower surfaces of the leaves by pushing spore-bearing conidiophores through the stomatal openings. Spores are produced on the upper leaf surface also or on plant debris left on the ground. The spores, when abundant, give a delicate gray or pale-lavender sheen to the affected areas. Sunken, tan-colored, elongated spots may occur on the stalks just before harvest that require heavy trimming and loss of edible product. L. J. Klotz, at the Michigan Agricultural Experiment Station, found the best temperature for growth and germination of spores of Cercospora apii to be about 70 F. Spores survived desiccation on dried leaves more than 170 days.

They are well adapted to dissemination by air and can infect floral parts and grow into the seed coat. When the seed germinates, the fungus can attack the young cotyledons and from them pass to other leaflets. The life cycle takes 10 or 15 days. J. D. Wilson and I, working at the Ohio Agricultural Experiment Station, showed that the longer plants are left in the crowded seedbeds, the worse blight is apt to be later in the field.

An important leaf spot of carrot also is caused by a Cercospora, but it is nonpathogenic to celery as is the celery pathogen to carrot. Late blight, caused by the fungus Septoria apii var. graveolentis, can cause even more destruction than early blight in cool, wet seasons and on the later crop in the Northern States. It can attack any part of the plant above ground. As outer leaves and stalks turn dark and wither, the entire field may look scorched.

The fungus is seed-borne. It also overwinters on debris from a previous crop. It may get started in the seedbed where it forms small, circular, water-soaked spots on the leaves about one-sixteenth inch in diametre. In 10 or 20 days, the spots turn nearly black and become filled with many minute black dots, the fruiting bodies (pycnidia) of the fungus. Spores are formed in these closed, black, pear-shaped cups partly embedded in the plant. They are exuded during wet weather as gelatinous, snakelike tendrils, and require spattering raindrops rather than air currents for quick spread.

When celery is wet with dew, the clothes of workers can spread the fungus down the row. K. H. Lin, at the New York State College of Agriculture, found that the number of spores in a single pycnidium a structure no larger than the dot over this i varied between 1,448 and 5,493, with a mean for nine pycnidia of 3675. Dr. Wilson and I found more than 2,000 spots on untreated plants, with an average of 56 pycnidia per spot; potentially, therefore, half a billion spores can be produced on one plant. Germinating Septoria spores can penetrate directly through the epidermis as well as through stomata and as readily on the upper surface of leaves as on the lower surface, although there are only about one-third as many stomata on the upper surface.

The bacterial leaf spot is caused by a soil-inhabiting bacterium, Pseudomonas apii. In the Lake States and New Jersey it sometimes gives trouble on outer leaves in hot, humid weather. It makes small, circular, rusty-reddish-brown spots up to one-eighth inch in diametre; sometimes they have pale-yellow borders. The spots remain smaller than early blight spots and are a darker brown. They differ from septoria late blight in lacking the black pycnidia within.

The disease is not seed-borne but it often gets its start in the seedbed. It is most troublesome on the crop maturing in August and September. It was the first bacterial plant disease to be controlled by dusting with a fungicide, in 1922. Control measures for all three celery leaf spots are practically the same. In the absence of resistant varieties, growers have relied heavily upon the use of copper fungicides in the field ever since the work of B. D. Halsted in New Jersey in 1891.

No fungicide has exceeded bordeaux mixture in effectiveness, but the residues it leaves and the inconvenience of preparing it (compared with the low-soluble coppers, such as basic copper sulfate, copper oxide, and copper oxichloride, and the organic fungicides) have led more and more growers to abandon bordeaux. The use of disease-free seed and a 2- or 3-year rotation, to eliminate the two important sources of primary inoculum, have helped many growers. Because the Septoria fungus embedded in the seed coat usually dies within 2 years and the seed retains its vitality 3 to 6 years, many growers buy their seed in advance or ask for 2-year-old seed when it is available.

In Bermuda the blight problem was greatly reduced by microscopic examination of samples of all imported seed and rejection at port of entry of all seed lots showing pycnidia. Fresh seed can be treated in various ways to kill the pathogens that cause early and late blights: A dip in hot water for 30 minutes at 118 to 120 F.; a dip in formaldehyde solution (1 to 300) for 3 hours at room temperature, followed by a rinse; a preliminary soak in tepid water for 30 minutes, followed by a dip in mercury bichloride solution (1 to 1,000) for 5 minutes and by a 15-minute rinse. The hot-water treatment is the best. Dusting or spraying seedbeds with a copper fungicide two to four times (first practiced by growers in New York and Ohio) is an economical and effective way of reducing and delaying the onset of all three blights. It has been widely used. If treatment of seed and seedbeds does not eliminate the blights, lack of rotation might be to blame. Growers who cannot rotate crops can hold the diseases in check by spraying or dusting in the field. Local practices respecting materials and methods of field spraying and dusting vary a great deal. They depend upon variations in climate, on chances of losses at different times of the year, on growers' preferences, and on the extent of local experimental testing of the newer fungicides. Bacterial blight can be controlled with a dust of 20 parts copper and 80 parts lime. That was a standard treatment among muck growers in New York between 1924 and 1935. But its high lime content and the need to apply it when plants are wet has led to its gradual abandonment in favour of the low-soluble coppers. The modern trend is towards materials that do not clog nozzles very much and leave no unsightly residues.

This change was made possible by general adoption of the seed and seedbed practices I mentioned, which give better control of blights in the seedbed and hence a reduced amount of blight in the field. The low-soluble coppers have given fair satisfaction where disease potential was not too high, although the rate at which they wash and weather off has made it wise to shorten the interval between applications to 5 or 6 days on many farms, even to 3 or 4 days in southern Florida. Heavy fertilisation, the use of mulches between rows, side dressing with nitrogen, and adequate irrigation to keep celery growing rapidly are enough in some cool seasons up north to give growers a satisfactory crop despite early blight.

But those practices are not so reliable in a hot, dry summer, on the early crop maturing in August, or on the late crop in Florida, which matures in April when temperatures are rising. Nabam has been widely used in the Everglades in Florida. Ferbam, ziram, and the coppers have proved more suitable to the upland soils of the Sanford area.

In California the most satisfactory and inexpensive programme against late blight includes spraying with 3-3-50 bordeaux mixture (3 pounds copper sulfate, 3 pounds lime, 50 gallons of water); in some localities zineb and the fixed coppers are preferred even though they adhere less well to the plant. New York growers are giving up copper lime dust for the low-soluble coppers. A few use Dithane in liquid form. In most years control is satisfactory, but in some seasons many have been urged to go back to bordeaux mixture or to an alternate schedule of bordeaux and ziram, zineb, or captan.

In Massachusetts growers prefer a low-soluble copper to bordeaux, but the organics, zineb and ziram, if applied more often, are finding favour from the standpoint of safety and effectiveness. Zineb is generally considered more effective than ziram against Septoria. Growers in Colorado have started using zineb sprays. Ziram and tribasic copper sprays have been favoured in Oregon. Extensive tests of fungicides in Michigan culminated in the development of a yellow cuprocide-sulfur-talc dust, 7-30-63, which is easier to use than liquid bordeaux, possesses less lime than 20-80 copper-lime dust, is more lasting than other low-soluble copper dusts, and has better flowing and keeping qualities. Breeding For Resistance to celery blight has been made possible by the use of blight-resistant foreign plants brought in by the Eastern States Farmers' Exchange and the Department of Agriculture. The first was a hollow-stalked Danish celery, which was crossed in 1937 with Golden Self Blanching and given to plant breeders at Cornell University, who crossed it to their new yellows-resistant variety, Cornell 19, in 1940. Progenies were later crossed with Cornell 6 to improve quality.

Further testing and selecting in New York and Florida resulted in the release in 1951 of Emerson Pascal, the first celery variety fairly resistant to early and late blights and highly resistant to fusarium yellows. In 1940 and 1941, G. R. Townsend, then a pathologist at the Florida Everglades Experiment Station, discovered Cercospora resistance in a number of plants grown from seed that was brought from Turkey by the Department of Agriculture. The plants resembled celeriac. Because he was unable to induce them to set seed, he shipped seven plants to Cornell, where R. A. Emerson, after two more years, finally induced two of them to flower.

Emerson made reciprocal crosses between Cornell 19 and P. E. I. (for Plant Exploration and Introduction, Department of Agriculture) 115557. Workers began a programme whereby progenies highly resistant to both blights were being selected from that cross in Florida during the winters and in New York during the summers.

As celery is a biennial, progress has been slow, but facts of value to vegetable breeders have emerged: Blight resistance is due to more than one gene. Hollow stalk and green colour are each dominant in the F_1 generation and are governed by a single factor, and hence are easily eliminated. Susceptibility to black heart and to magnesium deficiency in the soil are both hereditary factors that can be eliminated by breeding.

The yellows disease was next in importance to the leaf spots in most of the Northern States until resistant varieties were discovered and developed in Michigan, New York, and California. Ray Nelson, G. H. Coons, and L. C. Cochran, at Michigan Agricultural Experiment Station, found there are three distinct diseases in the sense that three separate strains of Fusarium may be involved.

Stunting of the plant, vascular discoloration, and crown and root rotting are common to all three, but leaf symptoms clearly distinguish two of them. Hot, dry weather is necessary to bring on a full expression of yellows. When the weather turns cool and wet, even infected plants may recover partly. The first symptom, a lagging in the growth rate, is sometimes seen even in the seedbed. In the field, plants lose their glossy appearance and there is a yellowing of the outer leaves between the veins. Fusarium apii form 1 is responsible for those symptoms. When temperatures are high, the entire plant, if not more than half-grown, may turn yellow in a few days. Often leaves of an infected plant when about to turn yellow become brittle. When they are crushed in the hands, they crackle like dead twigs.

That is particularly true of green varieties, which usually never turn yellow; that and stunting may be the only visible indication above ground of the disease in such varieties. Petioles of affected plants often develop a bitter taste.

In the second type of yellows the earliest symptoms are a downward curling of the young heart leaves. Then the veins lose colour. In this type Fusarium apii var. pallidum or form 2 is at work, and the areas between the veins are the last to turn yellow. Stunting, root rotting, and browning of the vascular elements in the roots accompany above-ground symptoms in both instances.

The California variation of the yellows complex lacks symptoms of leaf yellowing, curling, and brittleness even in warm weather. It is primarily a. stunting, with secondary root rotting and primary vascular discoloration. The identity of the Fusarium that is responsible seems not to have been established. But it differs from the other two, which are more common causes of yellows in the Northeast.

In warm soil (77 to 85 F.), the symptoms below ground include brown discoloration and death of small secondary roots, vascular discoloration and dry rotting of the taproot and crown, and even splitting of the latter. Many plants die. The fungi responsible for fusarium yellows have been isolated from vascular tissue of all parts of affected plants from lower roots to upper leaf stalks. Fusarium apii form 1 develops a pink to purple colour on steamed rice, while F. apii var. pallidum or form 2 remains colorless.

Some writers put both to-ether under the name F. oxysporum form apii, but not this one. The fungi live in the soil for many years even in the absence of celery. Green varieties generally are much less susceptible. Many of them can be grown on infested land without fear of much loss.

There is also a difference in susceptibility of different strains of the same variety as well as a difference in apparent behaviour of one variety in different seasons. The most extensive testing of varieties has been done in Michigan by Ray Nelson and L. C. Cochran and in Ohio by J. D. Wilson.

The following lists of varieties resistant and susceptible to yellows are based partly on their findings:

- *Yellow Varieties More or less resistant*: Michigan Golden, Michigan Green Gold, Cornell 19, Cornell 6, Morse's Masterpiece, Florida Golden, Tall Golden Pascal, Plume, Golden 99, Golden Pascal, Emerson Pascal, and some strains of Wonderful.
- Moderately susceptible: Wonderful, Golden Plume (some strains), Kilgore's Pride, Golden Prise, Kilgore's Pearl Special, Sneck's Florida Golden, Early Fortune, Superplane, and Paris Golden.
- *Very susceptible*: Golden Self Blanching, Wonderful (some strains), Early Fortune (some strains), Meisch's Special, Hoover's Special, Gunson's Special, Golden Phenomenal, Golden 14, Golden Detroit, and Golden Plume (some strains).
- *Green Varieties Highly resistant*: Curly Leaf Easy Blanching, Pride of the Market, Full-heart Easy Blanching, Winter King, Autumn King, Woodruff's Beauty, Sweetheart, Crispheart, Krispgreen, Holmes' Crisp, Earligreen, and Newark Market. Moderately susceptible: Winter King, Utah, Fordhook, Columbia, Epicure, Pascal, Emperor, Newark Market, Crystal jumbo, and Winter Queen.
- *Very susceptible*: White Plume, Houser, and Paragon.

Some varieties are classified as resistant at one time or by one worker and susceptible at another time or by another worker. T. C. Ryker, who grew plants in infested soils held at different temperatures at the University of Wisconsin, pointed out a possible reason for this discrepancy.

He found some varieties or strains of a variety are resistant in soils up to a temperature of 79 F., above which they are susceptible. Early planting in the Northern States therefore enables a fairly susceptible variety to make good growth before soil temperatures become high enough to induce fungus infection. In wet, cool seasons infection may be delayed until the crop is made.

Control of celery yellows has been accomplished in Michigan by the selection of individual plants that grew well in fields, where nearly everything else died of the disease, and propagating directly from them. That was done between 1919 and 1926 with selections from a field of Dwarf Golden Self Blanching. The variety released was named Michigan Golden. A similar process of field selection and multiplication from a field of Tall Golden Self Blanching resulted in the Michigan Golden Tall strain between 1930 and 1933. Both were yellow varieties, but in 1951 a green variety, Michigan State Green Gold, was released. It was the result of hybridising the Downing Strain of Fordhook with the tall strain of Michigan Golden. It has a light-green stalk and is widely grown in Michigan and elsewhere.

At Cornell University in 1933 Swarn Singh crossed Utah, a moderately resistant green variety, and Golden Self Blanching, a popular susceptible yellow one. His work culminated in the development of two new varieties, Cornell 19 and Cornell 6, both self-blanching and highly resistant to the eastern strains of yellows. By crossing the blight-resistant strain described under early blight with each of these in succession, New York workers developed Emerson Pascal, a variety resistant to both diseases.

It was put on the market in 1951. Phobia root rot has occurred at times in the Northeast. The fungus that causes it, Phoma apiicola, can attack other members of the Umbelliferae, such as carrot, parsnip, parsley, and caraway, but not hemlock or dill. It can live over in the soil on plant debris and occasionally on the seed.

Its first appearance often is in the seedbed, where it causes stunting, yellowing of outer leaves, brown rotting of roots, and sometimes death of young plants. The fungus fruits on the roots and crown; spores from a diseased seedling thus can be spread at transplanting time to many other plants if they are pulled and soaked in water before they are set in the field.

The fungus usually is confined to the crown, but sometimes it grows up into the outer leaf stalks far enough to turn them a dark bluish-green colour and cause them to break over. Crown lesions take on a dark, brown, rough, scurfy appearance and frequently crack open in later stages. Plants may be killed in the field, although usually they linger on. The fungus prefers a comparatively low temperature, 61 to 65 F., and requires oxygen and moisture for its most rapid development. Therefore it causes trouble principally on the spring crop and sometimes the late fall crop when temperatures and moisture are most favourable.

The pycnidia of the fungus, filled with very small one-celled spores, may be found partly embedded in the root lesions at any time. The fungus can occasionally become embedded in the seed coat, and it has been introduced by this means into new territories where it caused bad local seedling infections. Free spores once extruded from their pycnidia, however, cannot survive more than 30 days on the surface of seeds at room temperature. No varieties are known to be resistant, although White Plume, Giant Pascal, and Easy Bleaching seem less susceptible. The fungus is not adapted to the climate of California, where most of our seed is grown. It has not been found in Florida, where much of our winter celery is grown. Losses can be kept down by treating infected seed with hot water, sterilising infested seedbed soil, rotating crops, and destroying plants that show symptoms. It has been confused with one or another of the blights of celery and with brown check and cracked stem but differs from them in several important respects.

M. A. Smith and G. B. Ramsey, of the Department of Agriculture, have described the chief symptoms of brown spot as irregular, light-tan or reddish-brown, shallow lesions, which occur on any of the above-ground parts of the plant. The lesions may unite to form a scurfy brown streak all the way up the inside surface of the stalk. Transverse cracks may develop across the large lesions and the fungus fruits in the cracks as well as on the surface of stalk and leaf spots.

A certain amount of distortion of growth may occur. The fungus produces many small, elliptical or elongate, one- and two-celled spores on the surface of infested areas. Although the fungus was first found and is much worse on the Utah Pascal types of celery, it occurs also on the Golden Self Blanching types and has affected as high as 85 per cent of the plants in a field.

Growers are concerned about this disease for several reasons. Because its fungus is a rapid, heavy sporulator, it probably requires more frequent applications of fungicides if satisfactory fungicides can be found. Because it often attacks inner stalks and heart leaves, the plant cannot always be trimmed down to a sightly, marketable product. Furthermore, plants from diseased fields, which look fairly free at harvest, when packaged at the wash house directly from the field for retail markets may develop unsightly reddish-brown freckles in transit or in storage.

The spores of Cephalosporium apii germinate best between 68 and 75 F. and not at all above 90 or below 45 . The fungus grows best at 75. Bordeaux mixture is more effective than wettable sulfur in inhibiting spore germination in the laboratory, but control measures have not been worked out in the field. Growers in New York up to 1952 failed to get very good control with either a low-soluble copper or a liquid Dithane spray applied as for blight control.

First attempts to isolate Cephalosporium from seed have failed. Fortunately a number of varieties are apparently resistant. These include Summer Pascal, Utah 52-70, Utah 15, a Department of Agriculture plant introduction 176789, and Tall Fordhook. Very susceptible varieties include Cornell 19, Golden Plume, Top Ten, Ten Grand, and Non Bolting Green Nos. 12 and 13 (from Hart and Vick), according to 1952 tests made at Cornell University by Ralph Segall. Stem checK, brown check, or adaxial crack stem is another new disease of celery which has caused heavy losses from coast to coast since the introduction, in 1943, of the splendid variety Utah 10B. It begins as light-tan, shallow, sunken, greasy-looking spots on the inner surface of leaf stalks after plants are half grown.

The spots or streaks turn dark brown and open up with a series of unsightly horizontal cracks. Sometimes typical symptoms of crack stem (known to be boron deficiency symptoms) occur on the outer ridges of the stalk but not always. Affected plants are not stunted; in fact, the disease seems to be worse where ample fertility increases growth rate. The disorder has been traced by P. A. Minges, J. T. Middleton, and other California workers to a deficiency of boron in the presence of excessive supplies of potash within the plant. Susceptible varieties seem unable to take up as much boron as needed to protect them.

This inability is probably an inherited character as certain strains of Utah, notably 10B and Utah Special, are very susceptible. Utah 16-5 and Top Ten are also moderately susceptible, while Utah 52-70, Utah 16-8, Utah 16 PC, and Summer Pascal, in field tests by the California scientists, were practically free from the disease. Control is only a matter of avoiding the use of susceptible strains, or if they are grown, then withholding of potash and excess nitrogen, or spraying with boron solution, or both, may be desirable.

At least nine virus diseases attack celery. Three or four of them are widespread and cause heavy losses. None is seed-borne. None remains in the soil after infected roots decay. Most of them have several common wild or cultivated plant hosts, which act as perennial reservoirs of infective virus.

Aphids are the usual vectors, but thrips carry spotted wilt and leafhoppers carry virus yellows. Cucumber mosaic virus, of which there are a number of strains, is common from coast to coast. The first symptoms are vein clearing and mottling of inner leaves. The most prominent symptoms develop about a month later and include stunting, fern-leaf growth of some leaflets, and raised, dark-green, blisterlike areas on others.

A closely related virus, causing southern celery mosaic, is established in Florida, Cuba, and Puerto Rico. It also occurs in the Northern States, where it caused heavy losses in 1950. Often buff-colored and translucent, sunken spots develop on outer petioles. F. L. Wellman correlated the spread of the disease in Florida with east winds and the spread of winged forms of the melon and cotton aphids. The corn leaf aphid and others also can transmit the virus. The disease has often been observed to start nearest to weeds or diseased economic plants. Eradication of Commelina nudiflora dayflower or dewflower in the Sanford area of Florida gave a large measure of control. The more than 140 host plants of the cucumber mosaic virus belong to more than 3o families, among them pokeweed, groundcherry, milkweed, and catnip.

Western celery mosaic occurs in California and Colorado. Its symptoms resemble those of southern celery mosaic, except that the leaf mottling is usually followed by necrotic spotting. On the petioles white spots or streaks develop instead of brown sunken ones. The disease became destructive in Los Angeles County, Calif., in fields where celery was grown continuously. It can be controlled if growers in an area observe a 3-month celery-free period beginning in September each year. The virus is restricted to umbelliferous hosts; celery, carrots, celeriac, dill, caraway, coriander and wild weed hosts are unimportant. At least 11 different aphids carry the virus.

The viruses that cause southern and western celery mosaics apparently belong to the nonpersistent group of viruses. They are easily transmitted mechanically. A vector does not retain for long the ability to transmit it. This ability may be lost during the first feeding or in less than 15 minutes. Spotted wilt occurs on celery chiefly in the cooler, coastal fog belt of California, where the virus attacks a wide range of truck crops and ornamentals. Symptoms of spotted wilt on celery are most pronounced on the outer rather than inner stalks and begin on older leaf blades as numerous small yellow spots, which later become necrotic. Internal pockets of dead brown tissue develop inside the petioles and become more or less visual from without as sunken brown patches, which may rot and result in death of the entire leaf. Plants are stunted and worthless.

The vectors of spotted wilt virus are the tiny thrips, Thrips tabaci and Frankliniella insularis, which must pick up the virus while still a nymph and in which a period of 5 to 9 days must elapse before the insect can transmit it. Once infective, the insect remains so throughout pupation, emergence as an adult, and often until death. The virus is not transmitted through eggs of the infective female. Control is through elimination of the host plants, including ornamentals, that harbor the virus in the off-season when celery is not being grown.

Spraying with one of the newer organic insecticides may be helpful in some cases. Tomatoes have been bred for resistance, but attempts to do that for celery have not been undertaken. Celery virus yellows is caused by the aster yellows virus so common on lettuce and carrots. It is not to be confused with fusarium yellows, which is caused by a fungus in the soil. The symptoms on celery include shortening, twisting, yellowing, and delicate mottling of inner petioles and leaves, later many new shoots develop, and there is some stunting and a general yellowing. In California, 23 to 100 days may elapse before symptoms begin showing after inoculation of celery.

The vector is the six-spotted leafhopper. That insect overwinters in the egg stage on winter barley and to some extent on native grasses. After developing into adults, the leafhoppers begin in June to migrate to more succulent host plants; they do some feeding on infected plants, among them wild carrot, plantain, dandelion, chicory, perennial sowthistle, and some species of wild aster. After the insect has picked up the virus, an incubation period of 10 days must elapse, during which time the virus multiplies within the vector.

The incubation period can be lengthened by heating the leafhoppers for periods up to 11 days at 91 F. After 12 days at this temperature they are no longer infective unless they feed again on a diseased plant. That accounts for the slower rate of spread of virus yellows during a hot summer.

Other celery virus diseases of less economic importance have been described in California under the names of western- cucumber mosaic, celery calico, celery yellow spot, crinkle leaf mosaic, and tobacco ring spot. Control of virus diseases depends upon doing away with the wild host plants harboring the virus, destruction of insect vectors, or use of resistant varieties. The first has been used in the control of the southern celery mosaic with considerable success in Florida. Destruction of weed hosts around celery seedbeds and plant houses in the north is being employed to reduce losses in the field. It is well known that when hay is cut, when pastures dry up, or when weeds mature, great migrations of aphids and leafhoppers take place when the weather is warm and the air calm.

R. C. Dickson in the Imperial Valley of California calculated some 40 million winged aphids on a mile front may pass in an hour and flights may keep up for many days or weeks. He found that an individual aphid may feed for less than a minute, before flying on to another host, so each plant may play host to many aphids a day. Weekly spray or dust programmes against the insects after the diseases appear have been disappointing in California, probably because the migrating females that bring in the viruses are not killed quickly enough to prevent their first feeding.

Celery is also subject to attack by several parasitic nematodes such as the root knot nematode (Meloidogyne species), the sting nematode (Belonolaimus gracilis), the stubby root nematode (Trichodourus species), and the awl nematode (Dolichodourus heterocephalus), as well as to damping-off disease (Pythium species), pink rot (Sclerotinia sclerotiorum), black crown rot (Centrospora acerina), rhizoctonia crater spot, a fusarium seedling root rot, a bacterial soft rot (Erwinia carotovora), bud failure, and two or more physiogenic diseases, such as black heart, crack stem (boron deficiency), and a pale-yellow mottle leaf (magnesium deficiency).

THE IMPORTANT DISEASES OF LETTUCE

American gardeners grow several types of lettuce the loose leaf varieties, cos or romaine, butterhead, and iceberg. All are in the botanical species Lactuca sativa. Most of the commercial acreage of lettuce is in the Southwestern States, where lettuce, most of it the iceberg type, is grown the year around and shipped in refrigerated cars to all parts of the country.

The lettuce plant has a compact mass of edible, tender leaves on a short stem. Its structure makes it very perishable. Harvesting and handling must be prompt. Lettuce is subject to a variety of diseases during its growth and its distribution to consumers. Seed decay and seedling blight, diseases that reduce plant stands, are caused by several soil-inhabiting fungi, such as Rhizoctonia solani and Pythium ultimum. The fungi attack the tender plants before and shortly after they emerge from the soil and cause the stems and roots to rot. Young plants may be attacked only at the soil surface.

If that happens, a short section of the stem rots and the seedling falls over. The injury is called damping-off. Older plants are seldom attacked. Conditions that reduce the rates of germination and emergence increase losses from seed and seedling rots in heavy, wet, poorly aerated soils, which tend to puddle. Losses in such soils are increased if seed is planted too deeply or if rain or irrigation water packs the soil about the seeds.

Lettuce seed germinates best at moderately cool temperatures in coarse, well-aerated soils. Losses from rots can be reduced, especially in heavy soils, by planting the seeds shallowly, and by irrigating before planting, or sub-irrigating. In districts where rain is likely to fall while the seedlings emerge, planting should be after, rather than before, a rain.

Seed decay can be reduced by dusting the seed with a fungicide, such as chloranil (Spergon), at the rate of 4 ounces to 100 pounds of seed; ferric dimethyl dithiocarbamate (Fermate), at 1 pound to 100 pounds of seed; or thiram (Arasan) at the rate of 1 pound to 100 pounds of seed. Yellow cuprous oxide (Cuprocide) is likely to injure the seed.

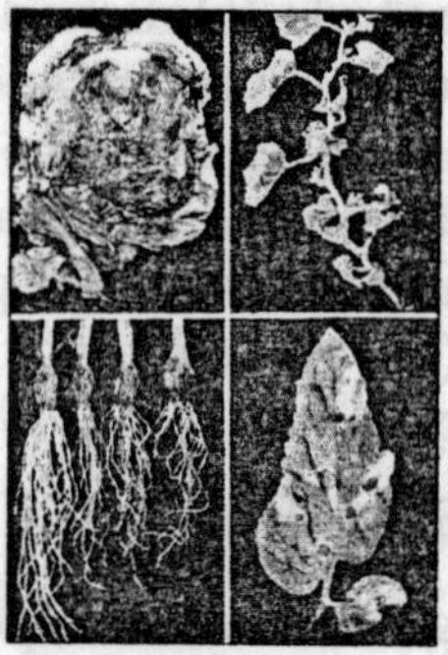

Upper left: Head lettuce affected with bacterial soft rot.
Upper right: Cantaloup vine attacked by the powdery mildew fungus. The disease is important in the Imperial Valley of California.
Lower left: Pink root, a disease of onion. One plant is healthy and three are diseased. Lower right: Alternaria leaf spot on tomato leaf.

Sclerotinia drop often causes severe losses in localities where rains come during the growing season, as in the Eastern and Central States. It is especially severe if lettuce and other susceptible hosts are grown repeatedly in the same soil. The disease causes occasional losses in Arizona and California, but it is seldom severe there because most of the commercial crop is grown on raised beds and harvested during rain-free periods.

Sclerotinia drop is caused by the soil-inhabiting fungi Sclerotinia sclerotiorum and S. minor. The fungi attack the leaves of older plants where they touch moist soil. They attack the stem at the moist axils of large leaves near the base of the plant. They cause a soft, watery rot, which spreads rapidly over the stem and leaf bases. The entire plant suddenly collapses. It becomes a soft, watery mass and then turns brown and dry.

A cottony fungus growth can be observed on the stem and leaf bases of plants that show early symptoms. Small black sclerotia of various shapes can be found in the decayed tissues as they become brown and dry. The sclerotic are resting bodies that enable the fungi to pass through periods of unfavourable weather. When moist, cool conditions favour their development, the sclerotia produce mushroom-like apothecia.

The apothecia produce innumerable spores which distribute the fungus to new host plants by wind, rain, irrigation water, and cultivating tools. The fungi can persist for long periods in the soil. They can also attack other crop plants, such as bean, cabbage, celery, eggplant, potato, and tomato. In places where the disease occurs on those crops, large populations of the fungi are built up in the soil and increasingly severe losses occur when susceptible crops follow susceptible crops in soils that harbor the pathogens.

Sclerotinia drop can be controlled in seedbed soils with steam pasteurisation or with chemicals: One part of commercial formalin in 50 parts of water applied at the dosage of 1 gallon to a square foot, or calcium cyanamide at the rate of 1,000 pounds an acre, 15 days before planting. Losses in seedbeds and greenhouses can be reduced by providing good ventilation and by using cultural methods that maintain a dry atmosphere and dry plant and soil surfaces.

In commercial lettuce fields it is best to grow the plants during rain-free periods and maintain a dry mulch around the bases of the plants. In the West the plants are grown on raised beds and irrigation water is applied in semi-permanent ditches. Irrigation water should be applied with care to prevent water-logging and to prevent wetting the soil surface in contact with the plants.

The building up of large populations of the fungi in the soil can be prevented by rotating lettuce and other susceptible crops with grains and other crops that are not attacked by Sclerotinia. No variety of iceberg lettuce is known to be resistant to Sclerotinia. The cos varieties are also susceptible but often are less severely injured than the iceberg varieties. The cos varieties have an upright habit of growth and present fewer moist infection courts to invading fungi.

Botrytis Rot and gray mold of lettuce are caused by the soil-inhabiting fungus, Botrytis cinerea. The plants usually develop brown, necrotic lesions on the stem near the soil line, on the bases of leaves near the soil line, or on leaves in contact with moist soil. Infection usually proceeds upward along the stem and inward through successive layers of leaves.

Hence, the rot often occurs in one side of the plant. In moist weather a gray mold appears on the dead tissues and on spots elsewhere on the leaves. Botrytis cinerea grows upon decaying vegetable matter in the soil and attacks numerous ornamental and vegetable crop plants during moist weather. It is often troublesome in greenhouses and occasionally causes losses in fields during wet, muggy weather.

Control measures for sclerotinia drop also apply to botrytis rot. Slime or bacterial rot of lettuce occurs in the field during warm, muggy weather; in transit, in lettuce shipped without refrigeration; and in markets. It often is troublesome in Eastern and Central States. In the West it is seldom seen during cool weather but often occurs in spring.

Slime causes a wet, slimy decay of the large internal head leaves. The outer leaves and the small leaves at the centre of the head are usually not affected at first, and the plants often appear normal in the field. They can be detected by twisting the top of the head. The outer and inner firm tissues separate readily at the rotted leaves. Slime is caused by Erwinia carotovora, Pseudomonas viridilivida, P. marginalia, and other species of bacteria. The growth of those micro-organisms seems to start in dead tissues, such as those caused by tipburn. Once started, the rot spreads rapidly to healthy tissues, progressing most rapidly within affected leaves.

The disease occurs under conditions favourable for the development of tipburn. Losses from slime can be reduced by using the control measures recommended for tipburn. Losses in transit and in markets can be kept low by providing adequate refrigeration and ventilation. Downy Mildew of lettuce occurs throughout the year in the coastal lettuce districts in California, where it is most severe in winter. It is less severe in winter in the drier valleys of California and Arizona.

The disease also occurs on lettuce grown under glass elsewhere in the United States and is especially severe during cool, wet weather in winter and spring. The first symptoms of downy mildew in lettuce appear as scattered light-green to yellow areas on the upper surfaces of exposed leaves. Within a few days downy, white hyphae and conidia can be observed beneath the discolored spots. Old spots become brown. The fungus continues to grow under refrigeration and predisposes lettuce heads to decay by bacteria and other organisms in transit and storage. Symptoms of downy mildew are not spectacular. Secondary organisms may be identified as the sole causal agents of infections initiated by the downy mildew organism.

Downy mildew is caused by an obligate parasite, Bremia lactucae. If moisture is abundant, the fungus penetrates the exposed leaves and grows between the cells. Some hyphae penetrate host cells without killing them and form absorbing organs (haustoria). The fungus produces conidia on profusely branched hyphae that project into the air through stomata. The conidia are carried to new locations by wind and rain and initiate new infections. They enable the fungus to spread rapidly during moist weather.

The fungus that causes downy mildew produces thick-walled resting spores (oospores) within the host tissues. They are released when the host tissue disintegrates. They enable the fungus to survive freesing temperatures and other unfavourable conditions. The fungus grows only on lettuce and some of its wild relatives. It may over-winter on weeds or in lettuce debris and volunteers in cultivated fields. Serious losses can be avoided by growing lettuce in a dry atmosphere during rain-free periods, by avoiding excessive irrigation, and by practicing clean cultivation to eliminate lettuce crop residue, volunteers, and weeds. C. E. Yarwood, working at the University of California at Berkeley, reported that downy mildew on lettuce can be controlled in California fields with 0.2 per cent zineb (Parzate or Dithane Z-78) spray with a spreader.

However, F. A. Haasis and D. E. Ellis, at the North Carolina Agricultural Experiment Station, found that more potent 2.0 per cent zineb drenches at seeding and at weekly intervals thereafter were required to control downy mildew on lettuce in propagation frames in North Carolina. When conditions are very favourable for downy mildew, control is difficult and expensive. Besides, it is best to avoid the use of fungicides on plants such as lettuce, the edible part of which consists of a mass of leaves.

The varieties Imperial 44, Imperial 152, Imperial 410, Imperial 456, Imperial 615, Imperial 847, Imperial 850, and Great Lakes are resistant to some of the races of downy mildew. These varieties are protected from injury in localities where only those races of downy mildew occur. Numerous races of B. laclucae exist. Some can attack all the varieties mentioned.

Downy mildew has been observed on all of them at some places. Resistance to the different races is very specific; that is, resistance to one race of Bremia does not protect a variety from attack by other races. The value of a variety in controlling downy mildew at any specific location can best be determined by trying it out in that location. Mosaic of lettuce causes losses in all parts of the United States. Its importance may not be recognised because the plants continue to grow and produce heads.

Losses are often attributed to unknown causes. A filterable virus, harbored in the seed and transferred from one plant to another by aphid vectors, causes the disease. It attacks plants of all ages. Plants with seed-borne infection and those attacked while young are dwarfed and exhibit leaves that are mottled with irregular, pale-green or yellow areas. Occasional plants are uniformly pale, yellowish green. The leaves often have excessively ruffled margins and may be distorted. They are usually more prostrate than those of healthy plants.

The plants may fail to form heads or they may produce small, loose heads of poor quality. Plants attacked when the heads are developing show mild symptoms, often on one side of the plant. The heads may be irregular in shape because of the unequal growth of diseased and healthy tissues. Mosaic plants produce seed stalks with mottled and distorted leaves. The plants lack vigour and produce little seed. The lettuce mosaic virus overwinters in the seeds of lettuce and, perhaps, in the weed hosts, sowthistle and groundsel. Usually less than 1 per cent of the seedlings are infected so that the disease causes little damage unless the virus is transmitted from the infected seedlings or from infected weeds to neighbouring plants by aphids.

The disease therefore causes little damage in winter lettuce. It often causes severe losses in summer lettuce in places where aphids are abundant while the plants are small. Seed-borne infection can be reduced by producing seed in fields free from wild Lactuca, Sonchus, Senecio, and other weeds and isolated from other lettuce. The fields should be rogued while the plants are small to destroy plants that were infected from the seed.

R. G. Grogan and his associates, working at Davis and Salinas, Calif.. demonstrated that mosaic can be controlled in market-production fields by the use of-mosaic-free seed. Other control measures include isolation from other lettuce and from weeds that harbor mosaic or the insect vector, the reduction of insect populations with insecticides, and the prompt destruction of crop residue in harvested fields.

No variety of the iceberg type of head lettuce is known to be resistant to mosaic. The mosaic-resistant Parris Island variety of cos was made available to growers by the South Carolina Agricultural Experiment Station in 1951. Aster yellows in lettuce is caused by a filterable virus, which leafhoppers transfer from one plant to another. The disease is widely distributed in the United States and causes severe losses in summer lettuce, especially in the East. Aster yellows has limited the expansion of lettuce production centres in the East and elsewhere. Most of the important production centres are in parts of California and Arizona where the disease seldom occurs.

Young lettuce plants affected by aster yellows have some curled and yellow or white leaves. Small brown Spots of dried latex may occur along the margins of the leaves. The leaf veins may be more translucent than those of normal plants. The whole plant may be affected or the infection may be limited to one part. Plants attacked after heading have twisted and dwarfed heart leaves and do not become firm. The axillary shoots often grow before the head is mature.

The flowering branches are chlorotic and slender and often bend down under their own weight. The flower heads on such branches are commonly dwarfed and may be distorted. The few seeds they produce are dwarfed and fail to grow. Aster yellows in lettuce is caused by the aster yellows virus, Chlorogenus callistephi. It attacks many cultivated plants and weeds. The California strain of the virus differs from strains found elsewhere in the United States in its ability to attack certain host plants, such as celery and zinnia, which are immune to other strains.

The virus is transmitted from one host plant to another by Macrosteles divisus and certain other leafhoppers. It can be transmitted by grafting but not by mechanical means. It is not transmitted by aphids. The virus commonly passes the winter in such perennial weed hosts as the common plantain. M. B. Linn, working at the Cornell University Agricultural Experiment Station, presented evidence that the vector does not pass the winter in the insect host or in cultivated fields and bordering weed areas in New York. His observations suggest that the insect over-winters in a milder climate some distance from the cultivated areas. The virus is apparently picked up from perennial hosts as the insects migrate into the fields.

A disease of this sort, with a wide host range and transmitted by migrating insects, is difficult to control. For the country as a whole, the disease is controlled by growing most of the commercial crop in districts free from the disease during the lettuce-production season. The best potential method of control in areas where aster yellows is troublesome is the development of varieties of lettuce resistant to the disease. That is not easy. Ross C. Thompson, working at the Plant Industry Station at Beltsville, failed to find immunity or potent resistance in cultivated lettuce (Lactuca sativa) or in related species that can be crossed readily with it. Certain collections of L. serriola and L. saligna exhibited marked ability to escape infection.

Those species may have value in breeding for resistance to aster yellows. L. tatarica, L. bourgaei, and L. marschallii appeared to be immune from aster yellows. Those species are cross-sterile with L. sativa. Some means must be found to cross them with cultivated lettuce before their immunity can be exploited. Until resistant varieties of lettuce have been developed, it will be necessary to use other methods to control this disease in areas where it is troublesome.

Dr. Linn recommended protection of seedlings from leafhoppers by cloth or metal screens, isolation of lettuce field plantings from weed areas and other yellows-susceptible crops by 200 feet or more; and weekly applications, from transplanting until 10 days before harvest, of either pyrethrum-sulfur dust (0.15 per cent pyrethrum) or rotenone-sulfur dust (1.0 per cent rotenone).

P. H. Bowser, working at the Michigan Agricultural Experiment Station, recommended the use of a DDT-treated trap area for migrating hoppers around lettuce fields in Michigan. If successful, this method would escape the residue hazard problem. Big vein of lettuce is caused by a soil-borne virus. It occurs wherever lettuce is grown. It is more prevalent in the summer districts than in the winter districts. Big vein causes severe losses in some fields, especially where lettuce is grown repeatedly for several successive seasons. The importance of big vein may often be unrecognised because the plants do not die but continue to grow and produce heads. Losses result from reduced yields of salable heads and poor quality.

The first symptom of big vein, observed in plants with five or more leaves, is a slight yellow discoloration along the veins. The yellowing of veins becomes more pronounced and the entire leaf becomes thickened and crinkled. All leaves of the plant are affected. No killing of tissues occurs. The plant continues to grow and forms a head. The symptoms fade as the head matures, but the head is smaller and less firm than normal heads and the leaves have poor flavour and texture. Although the above-ground parts of the plants exhibit marked symptoms, they do not contain the virus. The virus is limited to the roots, in which it multiplies. The amount of virus in infested soil is greatly increased when lettuce is grown. The big vein virus persists for long periods in the soil, but not in dried lettuce roots. The reason for this paradox is not clear.

The big vein virus can penetrate lettuce roots without the aid of nematodes or other known vectors. It is possible, however, that root aphids, nematodes, or other organisms act as vectors in natural infections. No variety of lettuce is known to be resistant to big vein. Big vein can be controlled in greenhouse and seedbed soils by treatment with heat, with D D or chloropicrin at the dosage of 0.46 milliliters to a gallon of soil, or with formaldehyde (1.6 per cent aqueous) at the dosage of 1 quart to a square foot. Big vein can be controlled in the field by growing lettuce in virus-free soil. Losses can be held in check in infested soils by rotation of lettuce with other crops that do not harbor the virus. Sporadic infected plants in otherwise clean fields should be removed carefully and destroyed by burning. The infested soil can be treated locally with D D, chloropicrin, or formalin. Saul Rich, working in Connecticut, recommended a mixture of 1 part commercial chloropicrin with 3 parts of xylene applied at the rate of 500 pounds an acre for the treatment of entire fields.

Brown blight of lettuce is a soil-borne disease of undetermined cause. It is of little importance at present, but it has considerable historical importance. It was a serious disease in California and Arizona from 1917 or earlier until its conquest through the use of resistant varieties developed by the Department of Agriculture and the

University of California. Nearly all of the commercial varieties now grown in California and Arizona are resistant to brown blight. The conquest of brown blight of lettuce by Ivan C. Jagger and Thomas W. Whitaker and their associates had beneficial effects on the lettuce industry and on the quality of lettuce on American dinner tables. The varieties they developed were not only resistant to brown blight; they were also superior to older varieties in uniformity, firmness of head, adaptation to culture during different seasons, and high quality. They are widely grown in the United States and elsewhere, and they have served as high-quality parents in breeding programmes.

Brown blight t attacks lettuce plants after they have developed five to ten leaves. Seedlings do not show symptoms. Plants attacked while they are small develop small, light-yellow spots in the young, expanding leaves at the centre of the plant. The spots enlarge slightly and the leaf areas between them become yellowish green. Those leaves and all subsequent leaves arc reduced in sise and tend to lie flat on the ground. The diseased plants are discolored, small, and rosette-like. Finally, the leaves turn brown and dry from the bases upward. Many plants die before harvest.

Plants attacked after heads form first show brown, irregular, disconnected, sunken blotches and streaks in the frame leaves or the larger head leaves. The streaks are usually associated with the midrib and larger veins, but they may occur between veins or along smaller veins. They are usually confined to the leaves, but they may extend into the stem. Many leaves or only an occasional leaf may show symptoms. The heart leaves are usually free from symptoms, but they may become brown and moist in nearly mature plants. In early stages, the roots appear normal. In plants with advanced symptoms the roots may be discolored and have dead root tips. The cause of brown blight is unknown. It may be a soil-borne virus. Observations by R. G. Grogan, of the University of California at Davis, and me in Imperial Valley suggest that the disease may be caused by the accumulation of toxic concentrations of substances in the soil about the lettuce roots.

Brown blight can be controlled by growing the resistant varieties Imperial 17, Imperial 44, Imperial 152, Imperial 410, Imperial 456, Imperial 615, Imperial 847, Imperial 85o, and Great Lakes. Imperial 101 and most other varieties derived from those varieties are also resistant. Brown blight can be controlled in susceptible varieties by growing them in disease-free soil. Tipburn is characterised by the death and dark-brown discoloration of marginal bands of the larger head leaves. In non-heading varieties, the symptoms appear in the rapidly expanding large leaves as they approach maturity. The mature leaves and the very young leaves are less often injured. The first symptom of tipburn is usually the breakdown and brown discoloration of small spots of tissue near the edge of the leaf. The spots usually occur first between the larger veins. The spots increase in numbers and coalesce as the disease progresses until the entire marginal band of tissue is killed. The symptoms may appear on only one or two leaves or on most of the leaves in the head. Under very favourable conditions most of the leaves may be involved. The dead tissues remain dry and are confined to the marginal portions of the leaves unless they are invaded by microorganisms. Various bacteria and other fungi may grow in the dead tissues, producing slime. Rotting then proceeds until the entire head is involved.

Tipburn in lettuce appears to be caused by the accumulation of excessive respiratory products in the sensitive tissues during warm nights. Injury seldom occurs at temperatures below 65 F. or during daylight. Environmental conditions that favour rapid, succulent growth predispose the plants to tipburn injury excessive soil fertility, excessive soil moisture, and warm temperatures. Environmental conditions that favour rapid respiration and the accumulation of respiratory products in the large head leaves favour the development of tipburn warm night temperatures and high relative humidity.

Tipburn can be controlled by growing lettuce during seasons when night temperatures do not exceed 65 F. as the heads mature; growing lettuce in soils that do not favour excessively rapid succulent growth; limiting the amount of fertilizers, especially nitrogenous fertilizers; using irrigation water sparingly when the heads approach maturity; and using resistant varieties. The Department of Agriculture, in cooperation with several State agricultural experiment stations, has released several varieties of lettuce that are tolerant of conditions that cause tip-burn.

The most resistant varieties, Imperial 456, Progress, Great Lakes and some of its derivatives, and Alaska, are adapted to culture during the summer. The moderately resistant varieties, Imperial 410, Imperial 615, Imperial 847, Imperial 850, and Jade, are adapted to culture during the spring. Resistance to tipburn is not complete, and all varieties suffer tipburn when conditions are very favourable. The resistant varieties often show little or no tipburn, however, under conditions that render susceptible varieties worthless.

Redheart of lettuce, a physiological disease, is characterised by the chestnut brown discoloration and breakdown of the small, inner head leaves. The outer leaves may appear normal or they may develop numerous, elongate, brown pits on the midribs and veins and sometimes on the tissue between the veins. It often occurs in transcontinental shipments of lettuce, especially in spring. The cause appears to be lack of sufficient oxygen, which results from poor aeration or prolonged exposure to low temperatures during shipment and storage. It also results from bacterial rot of the outer leaves, which occurs in shipments without adequate refrigeration.

Redheart can be controlled by providing adequate aeration and prompt and continuous cooling of lettuce to 39 to 41 F. in shipping containers, refrigerator cars, and terminal storage, and by prompt movement of the produce from the grower to the consumer. Premature yellowing, rib blight, and several other obscure diseases occur in head lettuce in the field and during shipping and marketing. The diseases occur most commonly during the spring and appear to be caused by various physiological disturbances.

Premature yellowing is associated with poor development of the root system and the production of small, infirm heads. Losses result from reduction in yield and from poor quality. Yellowing in early spring lettuce in Arizona was found to be associated with poor aeration, excessive soil moisture, and salt accumulation in the root zone. This complex of adverse conditions results from soil compaction by farm machinery and untimely applications of irrigation water. Losses can be reduced by avoiding the use of heavy machinery, especially on wet soils, and by avoiding excessive applications of irrigation water.

Rib blight is characterised by the occurrence of a dark-brown, necrotic strip of tissue along the midvein near the centre of the leaf. This symptom occurs in only one or two or in several of the large head leaves. The plants often appear otherwise normal and the disease can be detected only by removing the outer leaves. Rib blight appears to be associated with rapid, succulent growth; the plants are often vigorous and dark green in colour. The disease predisposes heads to attack by bacteria and losses are often attributed to slime. The cause of rib blight is unknown. Excessive applications of fertilizers and irrigation water should be avoided they favour rapid, succulent growth and predispose lettuce to injury by rib blight, tipburn, and other physiological disorders.

CAULIFLOWER, CABBAGE, AND OTHERS

The cabbage tribe includes cabbage, cauliflower, broccoli, brussels sprouts, kohlrabi, kale, and collard. All have been derived from the leafy wild cabbage of Europe. They inter-cross readily with each other and with cabbage. Somewhat removed from them botanically is radish, which does not ordinarily cross with members of the cabbage tribe.

When it does, a rank-growing, sterile hybrid usually results. Turnip and rutabaga are the other two important vegetables in this group. They are known to hybridise with each other. Their hybrids, too, generally are sterile.

All these vegetables belong to the family of plants known as crucifers, which includes also Chinese cabbage, water cress, rape, the wild and cultivated mustards, and many weeds, such as shepherds-purse and pennycress. Some 60 diseases may affect one or more of the cruciferous vegetables. The most destructive are yellows, black rot, blackleg and dry rot, club-root, and mosaic. Yellows, a warm-weather malady, is most destructive throughout the Corn Belt and as far north as central Wisconsin and northwestern New York.

It is not important on the midwinter crop in the South but may be destructive on autumn-sown cabbage and on the part of the crop that grows into late spring. Yellows appears on young plants, in the seedbed or after transplanting, as a lifeless, yellowish-green colour of the foliage. Often the yellowing is more intense on one side of the leaf, and on it a warping develops. Lower leaves are affected first. As the disease progresses up the plant, the leaves become brown and brittle and drop prematurely.

The water vessels of leaf and stem turn brown. Severely diseased plants die early. Entire fields may be destroyed if susceptible varieties are used on infested soil. Yellows is especially severe on cabbage, a crop often grown in climates favourable for the disease. It is less common on other susceptible crops, which usually are grown in cooler regions. The causal fungus, Fusarium oxysporum f. conglutinans, produces white or cream-colored threads, which form a cottony mass on culture media.

On the threads are borne numerous microscopic spores, which propagate the fungus. The fungus can exist indefinitely in the soil. It is not seed-borne but is transported in transplants and in soil moved about by wind, water, implements, and animals. The fungus in the soil enters the plant through the young roots at or near their growing tips. Without noticeably injuring the outer parts of the root, it becomes established in the water vessels and progresses upward through them to the taproot and into the stem and leaves.

The fungus remains confined in the water vessels until the plant is dead. Then it grows to the surface and produces spores. How it acts on the plant is not known. It probably plugs the vessels and produces toxic materials, which cause discoloration of the bundles, yellowing of leaves, leaf drop, and death. Yellows was so destructive by 1910 that it became the subject of a research project initiated by L. R. Jones of the Wisconsin Agricultural Experiment Station. The programme was continued as a cooperative project with the Department of Agriculture. It was one of the first studies of the relationship between soil temperature and plant disease. It brought out that when the soil was kept at 65 F. or lower little or no disease developed in plants grown in infested soil.

As the temperature rose, the intensity of the disease increased up to about 81 . Then it declined. This temperature curve followed closely that of growth of the fungus in pure culture. The results explained why the disease increased in intensity as summer temperatures rose and also why it was severe in places where cabbage grew in warm months and was little noticed where the crop was grown in winter, All attempts at practical control by chemical treatment of soils were unsuccessful. Dr. Jones noticed, however, that in any field that was almost completely destroyed occasional healthy plants survived.

By growing seed from such plants and returning their progenies to infested soil, he got higher per centages of resistant plants in succeeding generations. After three generations of selection within the variety known as Hollander or Danish Ballhead, he obtained a strain sufficiently resistant to be of commercial value. It was released in 1916 under the name Wisconsin Hollander. It was still widely used on yellows-infested soil in 1953.

Two genetic types of resistance to yellows exist. One, known as type B, is found in Wisconsin Hollander. Wisconsin Hollander may show a considerable per centage of diseased plants in hot summers, but most of the plants are only mildly affected. In the parent variety, Hollander, and in some other susceptible varieties, another type of resistance was found later. A single dominant gene controls it. This resistance, known as type A, has two advantages over type B resistance. As it is controlled by one gene, it is readily fixed and transferred from one variety to another. It also is stable at constant soil temperatures as high as 78 F.

Because breeding lines may carry both types of resistance and the two are hard to tell apart in the field, a way „as developed to separate them in breeding programmes. Type A is expressed more strikingly than type B in young seedlings and in old plants. By making use of the soil temperature effects studied earlier, technicians discovered that if young seedlings are grown in heavily infested soils at 75 , those that carry only type B resistance succumb to yellows while those which carry the gene for type A resistance are not affected. By growing seedlings in metal pans in thermostatically controlled water baths, known as Wisconsin soil temperature tanks, plants that carry type A resistance can be sorted out. The test is accurate and can be applied quickly and in small space to thousands of plants. By selection and hybridisation, type A resistance has been incorporated in many types of cabbage. Yellcws-resistant varieties to suit most seasons and market requirements are available, but the work of breeding new varieties to meet changing requirements continues.

The chief yellows-resistant varieties now in use are Jersey Queen, pointed head, early; Wisconsin Golden Acre, round head, early; Resistant Detroit, round head, second early; Racine Market, round head, second early; Wisconsin Copenhagen, round head, midseason; Marion Market, round head, midseason; Globe, round head, midseason; All Head Select, flat head, midseason-, Improved Wisconsin All Seasons, flat head, late; Improved Wisconsin Ballhead, round head, late; Wisconsin Hollander, a semi-round head, very late; Bugner, semi-round head very late; Resistant Red Hollander, round head, late.

Black Rot, much like yellows in appearance, has been known since 1890 or so. It occurs on the same crops as yellows. It also affects turnip, rutabaga, radish, Chinese cabbage, rape, and other leafy crucifers. It occurs wherever those crops are grown in the United States except in places in the Pacific coast region where the crops are produced in dry seasons. The disease starts, often inconspicuously, at the water pores of the leaves. The organism progresses throughout the plant by way of the water vessels. The organism is known also to enter through the root system and through wounds made by chewing insects. Marginal leaf infection is followed by browning and dying of the tissue. The lesion often forms a V, pointed towards the midrib. In.the lesion the veinlets become black. As the organisms advance, the invaded vessels continue to turn black through leaf, stem, and heads of cabbage and through roots of fleshy crucifers.

Non-marginal yellowish places next appear on leaves; as they turn brown and drop prematurely the disease closely resembles yellows and is often confused with it on cabbage and other hosts susceptible to yellows. The chief points of distinction are: The veins in yellows tend to be brown and those in black rot are black; V-shaped lesions at the leaf margins are not so distinct in yellows as in black rot; soft rot of cabbage heads commonly follows black rot but seldom does in yellows. The causal organism, Xanthomonas campestris, is a motile bacterium, which produces yellowish growth on culture medium. It was studied intensively in the 1890's by Erwin F. Smith, of the Department of Agriculture, and H. L. Russell, of the Wisconsin Agricultural Experiment Station. They worked out some of the essential features of the disease cycle. They learned that the bacteria entered the plant chiefly through the hydathodes at the edges of leaves on which guttation drops commonly accumulate during periods of high humidity, especially during cool nights which follow warm days. The bacteria soon gain access to the extremities of water vessels in that way. Only rarely do the bacteria enter stomata, the "breathing" pores that open and close probably because in cabbage the substomatal chambers in leaves and stems seldom become filled with water in nature.

When that condition is provided artificially, however, abundant stomatal invasion can be secured. In the first leaves of cabbage seedlings, a few stomata cease to open and close and serve more or less as hydathodes. Charles Drechsler at the Wisconsin Agricultural Experiment Station showed that that was the chief point of initial entry of the plant by bacteria carried with the seed. The seed coat is carried above ground as the seed germinates. It remains attached to the cotyledon, where it provides the infectious bacteria that penetrate the stomata.

Infected seed has been regarded as a major source of inoculum, but the disease cycle was not fully understood until the work of Allyn Cook, R. H. Larson, and me, of the Department of Agriculture and the Wisconsin Agricultural Experiment Station. It had been observed repeatedly that black rot epidemics appeared suddenly in cabbage fields approaching maturity, although there was no obvious source of the bacteria other than the seed used the previous spring. Plants that become infected in the seedbed commonly lose the infected leaves early. External signs of the disease disappear, although some bacteria have advanced into the stem before the infected leaf or cotyledon dropped.

The plant may grow for weeks or months without any perceptible reduction in growth, but the bacteria become distributed more or less generally throughout the plant. At some later time, for reasons still not well understood, the bacteria multiply rapidly and cause many lesions along the leaf margins and in many plants in a field. The abundance of inoculum so produced may be spread rapidly in warm, moist weather, and the destructive disease epidemic is at hand. To control black rot, it is necessary first to practice a 3-year rotation with non-cruciferous crops to avoid infection by organisms in plant debris. The next remedy is to prevent introduction of inoculum with the seed. Liquid or dust chemical treatments are not effective in this case. Immersion of the seed for 30 minutes in water at 122 F. is effective against black rot germs in the seed. That is a standard seed treatment. It is unnecessary to treat seeds of cabbage, cauliflower, turnip, and rutabaga grown in the Pacific Coast States, where the dry spell in midsummer keeps the disease from spreading in seedbeds. Blackleg dry rot affects cabbage, cauliflower, broccoli, brussels sprouts, kohlrabi, kale, collard, Chinese cabbage, turnip, and rutabaga.

It is incited by a fungus, Phoma lingam, which like the organism of black rot, subsists on infected plant debris for 1 or 2 years in the soil and in infected seed. All parts of the plants above or below ground may be affected. Spots appear on leaves as inconspicuous, indefinite, pallid areas. They gradually become well defined with ashen-gray centres, in which innumerable black dots, much smaller than a pinhead, are scattered irregularly; they are the fruiting bodies of the fungus, within which innumerable spores are formed and exude to the surface only in moist weather.

On the cabbage stems similar spots occur. Often they have a purplish border. When they are near the soil line, they gradually extend below ground, where the fungus eventually destroys the fibrous root system. That causes the plant to wilt or to topple over as the head increases in weight. When plants are carried through the winter to produce tall, branching seed plants, stems, branches, and seed pods are similarly affected. Here the fungus gains access to the young seeds. It burrows into the seed coat, where it remains dormant without necessarily stopping germination.

When turnip and rutabaga crops are infected, similar leaf lesions appear, and dry rot cankers on the fleshy roots develop before harvest or appear later in storage where heavy losses may follow. The fungus of blackleg depends entirely upon dew and rain to promote discharge of spores and to spread them to uninfected plants. The disease starts from infection of young seedlings by the fungus from infected seed or from trash in the soil. Humid, rainy weather is essential for an epidemic and conversely in areas with dry growing seasons the disease is relatively rare. In Pacific coast areas the disease is so rare that seed produced there is considered safe without treatment. The only effective treatment, when it becomes necessary, is keeping the seed for 30 minutes in water at 122 F.

No resistant varieties are known, but the disease is controlled by 3-year rotation with non-cruciferous crops and the use of seed grown on the Pacific coast or treated with hot water. One should not plant crucifers in fields next to fields where crucifers were grown the year before, because surface-water may spread infected debris. Clubroot, a troublesome disease for centuries throughout Europe, is acute in many intensive crucifer-growing areas in our Northern States. Most crucifers, wild and cultivated, are susceptible and thus the mustards, shepherds-purse, and pennycress can serve as perpetuating hosts of the causal fungus, Plasmodiophora brassicae. As clubroot affects only the below-ground parts of the plant, it may run part of its course after infection without causing any noticeable symptoms above ground. Such signs, when they appear, are likely to be in the form of slowly reduced growth, sometimes temporary wilting, and occasionally premature death.

When affected plants are pulled, various types and stages of root enlargement and malformation are found. The club may consist of a fleshy enlarged root in the shape of a spindle or it may consist of a spheroid gall. When many infections occur close together, most of the root system is transformed into variously shaped malformations. The clubbed tissue after a time is invaded by soft-rotting organisms. If the plant is not protected by production of additional secondary roots it may now wilt and die or, depending on the extent of root damage, slow down to stunted, unproductive growth.

The club tissue is permeated by the fungus, which first stimulates the growth of the host and then goes over into production of innumerable spores, which are released to the soil as secondary organisms decay the roots. The next spring the spores germinate and produce motile bodies, which enter the young roots of crucifers. Since the fungus does not gain access to the seed, it is not seed-borne. It is transported in diseased roots and transplants and in soil in various ways. There is no evidence that it lives as a saprophyte, but infested soil is known to remain infested for 10 years or more. Infection is favoured by relatively high soil moisture.

Clubroot is hard to control. Alkaline soil is unfavourable to it. In places where the soil is naturally alkaline or can be made so by liming, the disease can be held in abeyance. Heavy liming, however, is not desirable for such crops as potatoes, which scab most readily on such soil. It is important to select noninfested soil for the seedbed. The disease may be reduced somewhat by dissolving 1 ounce of mercuric chloride (corrosive sublimate) in 16 gallons of the transplanting water. That is not a complete remedy, however, and can be applied only in small cultures.

Varieties of turnip that are somewhat resistant are used in Europe. The strains of the clubroot organism prevalent in the United States appear to be relatively mild on turnip and rutabaga, however, and the disease is thus of little import on those crops here. A number of viruses infect crucifers. Disease is commonly caused on cabbage by the combined effect of two. One is a strain of turnip virus 1 and the other a strain of cauliflower virus 1. The first causes mottling of the foliage.

The second causes a clearing or chlorosis along the veins of the leaves. Together they produce lifeless spots on leaves, stems, and pods, pronounced stunting, and premature defoliation. The viruses carry over from season to season in cruciferous weeds and in plants kept for seed production. They are both transmitted to healthy plants by the cabbage aphid. The mottle phase is most noticeable in warm weather and the vein-clearing phase is favoured by cool weather.

Mosaic caused great damage in the early 1940's in cabbage seed crops growing near Puget Sound. By 1942 the reduction in seed production was so great that special measures were taken to study and control the disease. The research men found that the intensity of the disease was brought about by the fact that in seed production the viruses infect the cabbage during the first season, winter over in such plants, and bring about stunting and sparse seed production the second year. Moreover, aphids leave the maturing seed plants in midsummer and migrate to seedbeds nearby and infect the new crop. After several seasons of this close cycle, the first-year crop was almost completely infected before going into the winter.

Control was accomplished by breaking the cycle. Seedbeds were removed to isolated areas, and the first-year crop was thus protected from infection. The seed plants in the second year were found to be relatively free from disease, and the yields of seed were normal. In other areas relatively successful control may be assured by moving the seedbed to locations where cabbage or other cruciferous crops have not been grown recently and where weeds have not become heavily infected with the viruses.

Varieties of cabbage differ in their resistance to mosaic. The resistant plants become infected as do susceptible plants, but they produce nearly normal growth despite the virus. A programme of improvement was started in the Department of Agriculture in cooperation with the Wisconsin Agricultural Experiment Station. By continuous selection of tolerant plants with yellows-resistant varieties, strains resistant to both diseases can be obtained. One such variety, Improved Wisconsin All Seasons, has been made available to growers.

HAZARDS TO ONIONS IN MANY AREAS

Onions are grown in every State: In the South, onions are an early spring crop, mostly for immediate shipment, and Yellow Bermuda, Excell, and Crystal Wax are the chief varieties. In the North, the earliest crop is grown from sets produced the previous season, and the major part of the acreage, much of it on muck soils, is sown in early spring and harvested in September and October. Strains of Yellow Globe,

Yellow Danvers, and Sweet Spanish predominate. White Globe and Red Globe also are planted. Onion sets are grown mainly in northeastern Illinois and southeastern Wisconsin and near Greeley, Colo.

The chief set varieties are Ebenezer, Yellow Strassburg, White Portugal, and Red Wethersfield. All northern- grown varieties are suited to winter storage. Onion seed is grown chiefly in California, Idaho, and Oregon. Garlic, chive, and shallot are grown from sets. Welsh onion and leek are grown from seed. Welsh onion, like shallot, produces many small bulblets and both are used chiefly as "green" onions. While many diseases affect onion and its close relatives garlic, chive, shallot, Welsh onion, and leek a few stand out as potential hazards, particularly to onions, in many areas.

Downy Mildew is most destructive in New York, Michigan, Louisiana, California, and Oregon in an average year. In unusually cool, rainy seasons it is a major disease in other Midwestern States and once in a while in Texas and Colorado. It affects onion, Welsh onion, leek, shallot, chive, and garlic.

It usually appears in midseason as yellowish spots on the upper half of leaves of the onions. The fungus fruits on the surface of the spot as a bluish-gray, fuzzy mildew when humidity is high. The spots increase rapidly if moist weather continues. Spores produced on the surface are carried widely by air currents to cause new infections. The tops die back. The advance of the disease increases with high humidity and declines in dry spells. Plants are seldom killed, but growth of bulbs is reduced and the bulb tissue is inclined to be spongy and of poor keeping quality.

When expanding stems in the seed crop are infected, uneven, stunted growth follows. Spots on one side cause the stem to bend in the direction of that side. As the seed top grows heavier, weakened stems break over and seed is light in weight and poor in germination. The causal fungus, Peronospora destructor, is spread by wind-borne, short-lived spores. In old leaves are formed winter spores, which remain viable in the soil until the next season.

The fungus threads mycelium may also live in the bulbs and sets. From them infected plants may arise when they are used for the seed crop or for an early bulb crop. Perennial onions may also become a source of summer spores in early spring. Summer spores are produced most abundantly during the night at about 55 F. They are spread during the day. As dew accumulates the following night, they germinate most rapidly at about the same temperature and penetrate the onion leaf or stem. Dew is required for germination and penetration. Windy weather favours spread of spores, but there is less dew then. That is why low muck areas with poor air drainage are the ones where onion mildew often appears first and causes greatest damage. Downy mildew diseases usually can be controlled by protective sprays or dusts. Onion downy mildew has been an exception; many experiments have been conducted on it, but to little avail. At the Michigan Agricultural Experiment Station, Ray Nelson in 1951 reported encouraging results with a mixture of Dithane Z-78 and sulfur applied as a dust. We expect improvements in that line of control as other fungicides are developed.

Localities with dry summers have little to fear from mildew. The production of onion seed therefore has increased in Idaho and declined in California and Oregon. A resistant variety, Calred, was developed by workers of the California Agricultural Experiment Station and the Department of Agriculture from a cross between a resistant strain of Italian Red and an Australian variety, Lord Howe Island. Its seed stalks are highly resistant and the leaves are moderately so. It is adapted to growing districts in California. We need to breed the resistant character into the more widely used varieties.

Neck Rot 0is one of the most serious of the storage rots of onion. It appears shortly after harvest as a softening of the scale tissue. It begins usually at the neck and occasionally at a wound elsewhere on the bulb. The affected tissue takes on a sunken, cooked appearance as it advances steadily down one or more of the bulb scales. There appear later on the decayed tissue grayish masses of fungus threads, which gradually form a compact mat on the surface. If conditions are even moderately moist, a gray, powdery mass builds up on the surface. Meanwhile the cooked-appearing zone advances while the fungus mass follows it down the scale. The scale gradually shrivels. If many scales are affected, the entire bulb dries down to a crisp mummy.

The causal fungi are three closely related species, of which Botrytis allii is the most widespread. The gray, powdery mass on the surface of the decaying scales consists of myriads of spores of the fungus, which are picked up readily by the lightest of air currents. They will live for some days or weeks, but not through the winter.

In the fungus mats there sometimes appear hard black bodies (sclerotia), about the sise of a barley kernel. They are made up of finely woven fungus threads, which can survive freesing winter weather.

When cull onions are dumped from warehouses in the spring, the sclerotia give rise to spore masses in moist weather. The spores are carried to onion fields by air currents. They do not infect the growing plant, but when i they germinate they grow saprophytically, principally on the oldest leaves which are sloughed off as the plant develops. As the plant matures it becomes susceptible at the neck. If the tops are cut while still green, the wounded neck is ideal for penetration and infection by the fungus.

The saprophytic stage of the fungus builds up most effectively in cool, moist seasons. If such weather persists into the harvest period, spores are most abundant and infection is greatest. If the crop matures in dry, warm weather, the build-up of spores is reduced to a minimum and the disease which follows is negligible. That is why neck rot is not a major disease in areas where the crop matures in dry climate, such as the Rio Grande Valley, central California, Utah, and Idaho. In the more humid upper Midwest and Northeast, neck rot varies from season to season, depending on the climate just before and during harvest.

With those facts in mind, growers there can do much to reduce neck rot. They should allow the bulbs to mature well before being topped. Bruising in harvest should be avoided. If there is adequate ventilation in storage, the disease does not spread very much.

Artificial drying at harvest and during storage reduces somewhat the early advance of neck rot, and many growers use it as a standard procedure. All varieties of onion are susceptible once penetration has taken place, but a great difference still exists between varieties. White varieties are most easily infected. They therefore need the most attention at harvest. Yellow and red varieties are more resistant, but one must be careful with them also, especially when weather favourable to neck rot prevails.

The mild varieties of all colours are more susceptible than pungent varieties of corresponding colours. It is therefore important that such types as Sweet Spanish be allowed to mature well and be given the best possible airing. They should be protected from rain and dew during the curing process. The best storage conditions for onions include a temperature of 32 F. or slightly above and a relative humidity of about 65 per cent.

On onion the disease becomes manifest in young seedlings and at any subsequent time in the growth period of the host. Abnormal yellowing of roots is commonly associated with pink root but is not necessarily a stage of the disease. Affected roots turn pink, shrivel, and die. As the plant sends out new roots, they in turn become diseased and functionless. That happens throughout the growing season. The affected plants are not usually killed, but the reduced food supply results in the formation of mere scallions or small bulbs.

The causal organism, Pyrenochaeta terrestris, is made up of many races, which vary in growth characters and in virulence upon onion. Black fruiting bodies smaller than a pinhead sometimes appear on the diseased onion roots. Within them myriads of microscopic spores are formed. They are not important in the spread and perpetuation of the fungus, because the fungus seems to depend upon fungus threads, which grow and persist more or less indefinitely in infected soil.

I know of no practical way to eradicate the fungus from infested soil. The development of resistant varieties is therefore important. Such a programme is under way in the Department of Agriculture and several State experiment stations. At the Wisconsin Agricultural Experiment Station, R. H. Larson has worked out a method of subjecting breeding progenies to pure cultures of the fungus as they grow in white quartz sand at controlled, constant temperatures. After 2 weeks of growth in a liquid medium, the fungus threads are chopped into fine particles, which are mixed in the clean, sterile sand. The sand is placed in shallow metal trays supported in tanks of water regulated at a constant temperature of 80 F. Seeds are sown in the sand, and after 28 days the diseased seedlings are discarded and the resistant ones are transplanted to soil and grown on to bulbs. Thousands of seedlings can be tested in a month in this way. Resistant individuals are then used for further breeding.

Yellow Bermuda is one of the most tolerant of the common varieties. Plant breeders have discovered that they can increase that level of resistance by rigid selection. Resistance is an hereditary character that can be transferred to other types by breeding.

Chives, the Nebuka type of Welsh onion, and Giant Musselberg variety of leek also have considerable natural resistance. The Evergreen shallot, developed at the Louisiana Agricultural Experiment Station, and the Beltsville Bunching, a non-bulbing onion derived from a cross between onion and Nebuka Welsh onion and introduced by H. A. Jones of the Department of Agriculture, are resistant types developed by selection and breeding.

Onion smut was first reported in 1869 in the Connecticut River Valley. By 1888 it was of great economic importance there on old onion soils. It has become an important disease in most of the onion-growing areas throughout the Northern States as far west as Oregon and central California. In the United States the disease has remained strictly a northern one, although very likely the fungus has been transported frequently to southern regions. The disease also occurs on leek and Welsh onion.

Smut appears on the first leaf (cotyledon) soon after it emerges above ground as a dark, slightly thickened area. If most of the first leaf or later leaves is involved, they are swollen and tend to bend downward. On plants starting to bulb, raised black blisters appear near the bases of the scales. Lesions may break open and expose black, powdery masses of spores. Most infected seedlings die within 3 or 4 weeks after they emerge from the ground. Some plants survive weakly until midseason or later, and occasional plants produce bulbs with lesions on the outer fleshy scale and in one or more underlying scales. The fungus does not produce a rot in storage, but affected bulbs may be more subject than healthy ones to invasion by storage-rotting fungi and bacteria.

The causal organism, Urocystis cepulae, as seen in the black smut pustules in the leaves, consists of microscopic spores, which can live for many years in the soil. When soil is infested it remains so for many years, although there is no evidence that the organism grows and multiplies in the soil. The spores are not ordinarily seed-borne but are transferred widely on diseased sets and plants and locally by wind and water-borne soil.

The onion seedling is susceptible to infection by fungus threads growing from the spores only during the early seedling stage. If the plant escapes infection until the first leaf has reached its full growth it will continue entirely free from disease. Healthy sets or transplants that are planted in infested soil grow with complete immunity from infection. The fungus is sensitive to high temperatures, and if the young seedlings are growing in soil with temperatures of 80 F. or above they escape infection, because the fungus is very inactive and the plant grows through its susceptible period more rapidly.

This phase of the disease cycle was worked out by the writer, L. R. Jones, and F. L. Wellman, of the Department of Agriculture and the Wisconsin Agricultural Experiment Station. We interpreted the results as explaining why the disease is of no importance in southern areas where seed is sown in late summer in very warm soil. A great deal of research has been carried out on resistance of onion and its relatives to smut. R. I. Evans, of the University of Wisconsin, found that as the first leaf grew it became gradually more and more resistant and even though the fungus continued to enter, it had less and less success in establishing itself.

Welsh onion, especially the Nebuka type, is more resistant than onion, because the tissue of its first leaf becomes incompatible to the fungus earlier and much more rapidly. Three of us at the Wisconsin Agricultural Experiment Station and the Department of Agriculture studied resistance in crosses between onion and Welsh onion. Resistance was highly dominant and the hybrid scallion onion, Beltsville Bunching, has nearly as high resistance as Welsh onion. In 1952 we had not been able to introduce the resistance of Welsh onion to the bulb-onion type, by backcrossing the hybrid to onion. Control of onion smut centres around protection of the young seedlings from infection. About 50 years ago at the Ohio Agricultural Experiment Station. method was developed In which a stream of dilute formaldehyde was introduced in the furrow with the seed.

The method, known as the formaldehyde drip, has been a successful control measure ever since. The standard treatment is 1 pint of 37 to 40 per cent formaldehyde solution in 16 gallons of water applied at the rate of 1 gallon to 150 feet of row, or 1 pint in 8 gallons of water applied at the rate of 1 gallon to 300 feet of row. The treatment, however, is cumbersome, formaldehyde injury may occur in very dry soil, and heavy rains immediately after application may reduce its effectiveness.

CONTROL OF DISEASES OF POTATOES

The Irish potato is susceptible to many diseases. The fungi that might attack it range from the slime molds to the smuts and rusts. It is subject to several viruses of the yellows and mosaic groups. Such nonparasitic diseases as black heart, sunscald, freesing injury, and a malnutrition caused by deficiency in magnesium, potash, and boron may cause damage. Several nematode diseases have been found on it.

Unless effective methods of control are practiced, serious diseases, such as late blight, ring rot, and leaf roll, can cause the total loss of a crop. Late blight of potato is a downy mildew caused by a fungus, Phytophthora infestans. The mildew, or flour-like spots, usually on the lower surface of the leaf, distinguishes the disease from other leaf spots on the potato plant. Humid conditions favour it.

Despite its name, the first infection often occurs soon after the plants emerge when favourable moisture and temperature prevail. At 70 to 75 F. the fungus grows so fast inside the leaves that within a week after infection it causes dead spots one-half to 1 inch in diametre. The entire plant may be killed within 2 weeks. The brown discoloration of the foliage brings to mind the terms "blight" and "rust."

Late blight destroyed potato crops so often during the nineteenth century that it outranked other potato diseases and was called "the potato disease." In 1845 it was responsible for the potato famine in Ireland. The late blight fungus produces a mycelium, or moldlike growth, which invades and kills the plant tissue.

From the mycelium grow treelike sporangiophores, which bear sporangia, the agents involved in spreading infection. At about 70 and 90 per cent relative humidity, sporangia develop in 5 to 7 days after infection. In water at 72 to 77 , sporangia form germ tubes. At lower temperatures, 58 to 63, they develop swarmspores, or zoospores. The two methods of germination enable the fungus to adapt itself to a rather wide range of temperature. The higher temperature favours mycelial development, rapid invasion, and killing of the plant. The lower temperature favours formation of zoospores and rapid increase in inoculum to spread infection.

The fungus infects the potato tuber as well as the tops. Spores from infected tops are carried by rain to the tubers in the soil. Tubers also are readily infected if they are harvested before the blighted tops are killed. Infected tubers quickly show a brown discoloration, which changes to a purplish colour. The fungus usually invades the tuber to about one-fourth to one-half inch below the skin. At about 36 to 40 , the affected tubers persist in a dry rot condition. Higher temperatures favour secondary infection with bacteria, and the blighted tubers develop soft rot.

The organism is carried into the following season in infected tubers. When such tubers are planted, the fungus invades the shoots on which form the spores that infect the foliage. Another source of infection is the blighted tubers in the potato refuse or dump pile. Late blight appears on potato tops on waste piles a few weeks before the plants in the fields emerge, so that the potato tops in the fields become infected as soon as the plants emerge and before spraying begins.

The spread of late blight is favoured by humid conditions such as may prevail in the New England States, other regions along the Atlantic coast, the North Central States. With favourable moisture, it also appears in the Middle Western, Southern, and Pacific Coast States. It occurs in long periods of rain, fog, and dew, which favour fruiting, infection, and dissemination of the fungus.

Control measures are sanitation, protection, and the use of resistant varieties. Sanitation involves careful inspection and removal of blighted tubers from seed potatoes and elimination of piles of waste potatoes. Waste potatoes that have to be dumped should be given a dormancy or sprout-inhibiting treatment 0.25 per cent isopropyl ester of trichlorophenoxy acetic acid before dumping to prevent sprouting.

For many years a combination of copper sulfate and lime bordeaux mixture has been used to control late blight. Weekly spraying with it during the growing season effectively, although not absolutely, controls late blight even during epidemics. Copper-lime dusts control late blight if applied at rates of 25 to 30 pounds the acre at weekly intervals. Some of the more recently developed products like Dithane and Parzate are less toxic to foliage and therefore favour higher yields than when bordeaux mixture is used. Because these products are not generally so effective for control of late blight as bordeaux mixture, some growers change to bordeaux mixture during the latter part of the season when conditions favour an epidemic of late blight.

Because lime in bordeaux mixture increases transpiration to a harmful degree and because one-half the amount of lime used formerly in bordeaux mixture is satisfactory, many growers use a recommended formula such as 10-5-100-10 pounds of copper sulfate and 5 pounds of lime in 100 gallons of water. Search for resistant varieties began about the middle of the nineteenth century. Varying degrees of resist- ante appeared in different varieties, but no cultivated varieties manifested immunity from late blight. Sebago, 'Menominee, and Calrose, though not immune, are somewhat blight-resistant. The tubers in some varieties, such as Sebago, are more blight-resistant than the foliage.

Among certain wild species, such as Solanum demissum, some lines or strains are immune. Some lines of that species also are blight-susceptible. Empire, Ashworth, Chenango, Essex, Placid, Virgil, and Kennebec are varieties in which S. demissum was one of the parents and are immune to some races of late blight.

The development of blight-resistant varieties is a complicated procedure actually, the plant breeder has to develop resistance to several biological races that vary in ability to cause disease. A blight-immune variety may be immune to one race but susceptible to other races. Within the past decade, however, efforts to develop blight-immune varieties show that varieties can harbor immunity to more than one race of the fungus of late blight.

Common Scab, caused by Streptomyces scabies, is recognised by slightly raised spots or lesions of rough, corky tissue on the tuber. The lesions may be so numerous as to involve the entire surface. They may involve distinct russeting or deep scab. Scab lesions spoil the looks of the tuber and cause waste in peeling and reduction in grade.

The fungus lives in the soil and on infected tubers. It occurs in nearly every potato-growing region. Soil acidity, moisture, temperature, and aeration influence the development of scab. A soil reaction below pH 5.2 is unfavourable for most of the common scab races, although some strains are said to cause infection below pH 5. In general, the fungus is favoured at pH 5.2 to 7. Common scab develops at a wide range of temperatures, 50 to 85 , but it thrives best at about 70. Relatively high soil moisture tends to check the disease in some localities, but in other districts high moisture may favour scab. Lack of aeration inhibits development of the organism. The several races of the fungus vary in type and colour of mycelium, colour in the medium, and pathogenicity. Apparently the races result by mutation and some races mutate more than others. Treatment of seed tubers with disinfectants kills the scab fungus on the tuber, but it fails to control the disease if the treated tubers are planted in scab-infested soil. The use of 300 to 500 pounds of sulfur an acre reduces the severity of scab in some soils. The use of ammonium sulfate in fertilizer that increases soil acidity may inhibit the disease somewhat. None of the treatments has been entirely adequate for scab control, however.

The most promising method involves the development of scab-resistant varieties. Studies in Europe have disclosed that Jubel, Hindenburg, Ostragis, and Arnica are scab-resistant. In this country, scab resistance was noted in Russet Rural and Russet Burbank. Although the scab-resistant European varieties are not adapted to growing conditions in America, they do serve as resistant parents. From them the resistant Cayuga, Menominee, Ontario, and Seneca were derived. Investigations on scab resistance by the Department of Agriculture and State experiment stations are designed to develop varieties that are superior to them in scab resistance, quality, adaptability, and yield.

Verticillium Wilt, caused by Verticillium albo-atrum, induces wilting of the tops and vascular discoloration of the stems, tubers, and roots symptoms similar to those associated with other wilt-inciting fungi. Since V. albo-atrum is favoured by lower temperatures than are wilt-inciting Fusaria, it often occurs in the cool seed-potato areas. Verticillium wilt is recognised by flaccid, drooping leaves. Often only one stalk in the hill or a part of a stalk may be wilted. Later all the plants in the hill are involved and finally are killed.

Cross sections near the stem or stolon end of the tubers expose dark-colored vascular elements, the sap-conducting elements. The verticillium wilt fungus does not cause a disintegration or rot of the invaded tissues. Sometimes other organisms may enter the wilt-invaded parts, however. Verticillium wilt on potato was reported in Europe and America early in the twentieth century. It occurs in the seed-potato areas in the New England, North Central, and Northwestern States.

The wilt fungus is harbored in the tubers and persists in the soil. If conditions are favourable, wilt-free soil can be infested by wilt-infected seed potatoes. Attempts at getting wilt-free seed tubers by cutting off the discolored stem end of infected tubers have met with failure because fungus hyphae may penetrate beyond the discolored section of the tuber. Roguing wilted tops has reduced the amount of wilt of seed potatoes if the tops in the three hills on each side of the wilted hill also were removed. In some regions the wilt fungus persisted in the soil after a 2-year rotation, but infestation was practically eliminated after a 3- to 4-year rotation. Some varieties Menominee, Saranac, and Sequoia are more resistant than others. Blackleg, caused by Erwinia atroseptica, is recognised by an inky-black lesion on the base of the stem, which is a primary distinguishing character from similar soft rot caused by bacteria. The blackleg bacteria invade tops and tubers.

Infection may enter the stalks from the infected seed piece. Diseased tops manifest slight to severe dwarfing and somewhat rigid and rolled leaves that in the later stages yellow, wilt, and die. In very wet weather, the inky-black lesions at the base of the stalk may spread to most of the plant. The bacteria enter the new tubers through the stolons of a blackleg stalk and invade the vascular elements, as well as other tissues of the tuber. Affected tubers show soft rot, involving the entire tuber. Under less favourable conditions the decay is arrested so that only the tissues in the centre of the tubers are disintegrated.

In very moist weather, bacteria in the soil invade freshly cut or poorly healed seed pieces a possible explanation for the more general appearance of blackleg in wet than in dry seasons in some places.

Because cuts, bruises, and other injuries permit the entrance of rot-inciting organisms, tubers should be handled carefully to avoid bruising. Storage should be provided with favourable temperatures and humidity for healing over (or suberising) injured tissues. Removing from the seed potatoes all tubers showing rot and storing 9 cut seed potatoes immediately after cutting at about 70 F. and 80 per cent humidity to favour adequate healing of the cut surfaces have been effective control measures in some places.

We know of no varieties that are immune to blackleg, but some contract blackleg less easily than others. In 1947 Carl Stapp, a bacteriologist in Germany, reported that Carnea, Flava, Johanna, Priska, Robusta, Sickingen, and Starkeragis were more resistant to blackleg than the 56 other German varieties he included in his test. Ring Rot, caused by Corynebacterium sepedonicum, is a bacterial disease. It is recognised by wilt of the foliage and rot of the vascular ring of the tubers. Chlorosis, or yellowing, and marginal browning and wilting of the leaves are symptoms.

In the tubers the disease is detected by a light-yellow discoloration of the vascular elements, which break down and exude a cheesy bacterial and cellular ooze when a tuber is squeezed. When other micro-organisms enter, the lesions enlarge, and Soft rot of the tuber results. Cankers or fissures often appear on the surface of infected tubers in the later stages. Bacterial ring- rot was found in Maine in 1932, a year after it was reported in eastern Canada. By 1940 it was detected in most of the potato States. In some areas it caused severe losses. Ring rot apparently is not harbored in the soil, but infected tubers over-wintering in the soil may develop infected volunteer plants, which may serve as sources of infection. The disease is spread from infected to healthy tubers by the seed-cutting knife, planting machines, grading equipment, and contaminated hands, gloves, bags, baskets, barrels,. and bins that have come in contact with diseased potatoes.

Control involves prevention, sanitation, and the use of resistant varieties. Control of ring rot is primarily the responsibility of seed-potato growers, .who should propagate and maintain their own seed potatoes. Potatoes that are free from ring rot can be kept so if no infected seed potatoes are introduced from other sources. If a grower must bring in seed potatoes from other farms and localities, however, he will be safe if he plants them in separate fields for a few years or until he is absolutely certain that they are free from ring rot.

Seed-potato certifying agencies in Most States set up a rigid zero tolerance for ring rot in certified seed potatoes in Largely as a result of those standards, the spread and extent of ring rot has been reduced greatly, and growers of seed potatoes have access to an adequate supply of healthy potatoes even if they cannot grow their own seed stock. Sanitation involves selling for table stock the crop that harbors ring rot and disinfecting all equipment, machinery, containers, and bins that in any way were in contact with the diseased potatoes.

Bins, crates, barrels, and bags can be treated with copper sulfate (25 pounds to 100 gallons of water). Machinery can be treated with formaldehyde (1 part in 10). Ring-rot free seed potatoes, such are produced by foundation seed-potato growers or on State foundation seed-potato farms, should be planted. By following this clean-up programme many potato growers have eradicated ring rot from their farms. Although no variety is immune to ring rot, some varieties contract the disease less easily than others. Teton, introduced in 1946, is an example. Virus diseases of potatoes have been reported in every country. Leaf roll and latent mosaic apparently occur most widely. The nature and cause of virus diseases have been under investigation for more than 100 years.

Many investigators now regard them as filterable proteins that multiply in living cells. The virus diseases of potatoes can be classified as mosaics and yellows. Distinguishing characters of the mosaic group are mottling and the ability to be transmitted in the sap. A mark of the yellows group is chlorosis; it cannot be transmitted in sap. The mosaic group includes several viruses, some of which have several strains or races that differ in pathogenicity. The yellows group includes the viruses that cause leaf roll, witches'-broom, and aster yellows. As with most of the fungus diseases, the potato virus diseases are designated by common names that indicate the response or reaction of the plant to the virus. The reactions or symptoms are affected by temperature, host, and the strain or race of the virus. Mottling usually is absent at higher temperatures.

Roll of the leaves is favoured by relatively high temperatures. Some varieties manifest the effects of a virus; another variety might act as a symptomless carrier of the same virus. One race of a virus can cause a reaction in the same variety that differs from the reaction induced by another race. Most of the potato viruses are transmitted by sucking insects (aphids and leafhoppers) but some are spread also by chewing insects. Some viruses are spread by mere contact of diseased plants with healthy plants and by mechanical agents, such as the seed-potato cutting knife and planting machines.

Although several mosaic diseases incited by as many different viruses affect the potato, I shall describe only three latent mosaic, virus A, and vein banding mosaic, which occur in many potato regions. Latent mosaic, caused by virus X, is so called because many varieties of potatoes are symptomless carriers. Synonyms of the latent mosaic virus are potato virus X, potato-acronecrosis virus, simple mosaic virus, potato virus 16, Solanum virus 1, and Annulus dubius.

Latent mosaic virus is harbored by potato varieties more generally than any of the other potato mosaic viruses. Almost every plant in most of the older varieties harbors the virus. Many varieties manifest the milder strains of this mosaic by slightly light-green and rugose leaves when the weather is cool. Latent mosaic virus has several races or strains that induce a range of reactions, varying from weak or no apparent symptoms to severe necrosis. A plant harboring one race may be protected against infection from other races of this virus. More than one race may be harbored by a plant, suggesting mutation or simultaneous infection by more than one race.

Other plants in the nightshade family that are susceptible to latent mosaic are tomato, tobacco, pepper, jimsonweed, petunia, physalis, henbane, woody nightshade, and black nightshade. Jimsonweed is a good one for detecting different races of latent mosaic virus. It reacts to the races by manifesting no symptoms for the weaker races to severe necrosis for the more virulent races. Globe amaranth, or bachelor's button, which manifests red local lesions, also is a good host for detecting latent mosaic virus. Latent mosaic is easily transmitted mechanically. Contact of diseased with healthy plants and the cutting knife spread the disease. Some sucking insects, such as aphids, fail to transmit latent mosaic, but grasshoppers are reported to transmit it. Reductions in yield of 10 to 30 per cent are reported for the common races of latent mosaic. Virulent races can cut yields more than 50 per cent. Because most of the old varieties harbor latent mosaic, it is possible that yields from them are 10 or 15 per cent lower than the yields of healthy stock.

Virus A often induces slightly light-green and rugose leaves, symptoms that are similar to those induced by virus X, the cause of latent mosaic. The symptoms are favoured by cool temperatures and are masked at high temperatures. Some varieties merely show light-green foliage. Others develop dead spots in the foliage.

Although tobacco and some other species in the nightshade family are susceptible to virus A, the symptoms often are so like those induced by virus X that diagnosis is difficult. Virus A is transmitted by aphids. It can also be transmitted in the sap. It is transmitted mechanically less easily than virus X. Virus A apparently involves several races, which vary in degree of reaction they induce on their host plants.

Virus A reduces yields by 15 to 25 per cent in some varieties, apparently by causing plants to mature too early. It does not affect the set of tubers. Since most of the old varieties generally harbor virus X, they carry a composite infection after contracting the aphid-transmitted virus A. On many varieties the two viruses induce more pronounced symptoms than those of one of them, such as distinct mottling and crinkling. This composite infection is recognised as mild mosaic, a term first used to designate mildly affected plants, as distinguished from severely dwarfed, curled, rolled, and necrotic plants.

Vein banding mosaic, caused by virus Y, is distinguished by rugosity, or wrinkled surface of the leaf, and vein clearing in early stages and by necrotic spots and streaks and leaf drop in the later stages. The symptoms vary in different varieties, ranging from slight rugosity to severe dwarfing, curling, and necrosis. Synonyms for vein banding mosaic virus are potato virus Y, Solanum virus 2, and Marmor upsilon.

Vein banding mosaic occurs in leading potato areas of every country. In some places a high per centage of the crop is infected. Susceptible hosts, besides potato, are tomato, black nightshade, henbane, woody nightshade, petunia, Solanum nodiflorum, Datura innoxia, and tobacco. The several races of vein banding mosaic vary in virulence. Vein banding mosaic virus is readily transmitted by sap inoculation, grafting, and aphids. It is responsible for a greater reduction in yield than that caused by either virus X or virus A. Infected plants die prematurely; the result is undersised tubers. Usually less than half a crop is obtained from diseased Plants. Because the older potato varieties generally harbor virus X, they harbor two viruses after contracting vein handing mosaic virus (virus Y). This Composite infection is known as rugose mosaic, which often causes more severe host reactions than either one Of the viruses. On some varieties, however, the vein banding mosaic virus induces as severe reactions as the composite infection.

Leaf roll occurs in potato regions in every country. In some areas all potatoes are infected so that growers in such areas do not know their stock is infected. They suppose that leaf roll symptoms are characteristic of a healthy crop. Synonyms for the leaf roll virus are potato phloem necrosis virus, potato leaf roll virus, potato virus 1, Solanum virus 14, and Corium solani. Apparently the leaf roll virus involves more than one race. Other hosts are tomato, jimsonweed, woody nightshade, black nightshade, and ground-cherry.

Leaf roll has not been transmitted by sap inoculation, but is transmitted by tuber and stalk grafts. Aphids, especially the green peach aphid, Myzus persicae, are vectors. Heavy infestations of aphids, mild temperatures, and absence of diseases and parasites of aphids contribute to the extensive spread of leaf roll. Leaf roll is responsible for reduction in sise of plants and tubers, as well as in set of tubers. Reductions in yield might be as high as 50 per cent. In hot and dry weather, leaf roll infected plants produce chiefly culls.

Control of potato virus diseases involves the use of isolated tuber-unit seed plots or fields and the resistant varieties. Because insects transmit virus diseases, it is useless to attempt to grow healthy plants near potato fields that harbor the diseases. Healthy potatoes can be grown in fields that are surrounded by woods and located at a distance from other potato fields. The healthiest seed stock should be used for planting the isolated field, and it should be planted as soon as the soil is tillable. Planting the tubers in tuber units; planting seed pieces of the same tubers in consecutive order facilitates detection and removal of diseased plants. Weekly inspections, beginning when the plants are about 6 inches tall, to detect and remove diseased plants will help control the diseases.

The number of plants that contract diseases increases as the season advances. Some growers therefore practice early harvesting. That means harvesting seed potatoes about 10 weeks after planting. The most effective (but laborious) method is to pull the tops so that the tubers can be harvested later when storage conditions may be more favourable. Killing the tops with a weed killer frequently fails to kill the entire stalk that produces leaves, so that aphid infestation and virus transmission are encouraged unless the tubers are harvested a few days after the tops die. In tests in northern Maine in 1932 to 1947, two highly susceptible varieties remained free from virus diseases when harvested August 1, even though they had been grown within 500 feet of diseased potatoes. Samples of the varieties, left on the same plots until harvest on September 15, had contracted 90 to 95 per cent of mosaic and leaf roll in seasons when aphids were numerous. The tests show that susceptible potatoes can be maintained free from virus diseases in some seed-potato-producing areas, even with meagre isolation, if the seed potatoes are harvested early in the season before the dispersal of virus-carrying insects.

Seed-potato certification agencies in different States inspect and certify potatoes that meet their standards. To have potatoes inspected, the grower notifies the agency, which charges a small fee for the service. Certified seed potatoes, though not necessarily disease-free, usually are superior to non-certified potatoes for seed. Certification, more than any other factor, encourages growers to produce healthy seed potatoes. Potatoes vary in their resistance to virus X. Some varieties contract virus X in more than 50 per cent of the plants. Under like conditions other varieties become infected in less than 1 per cent of the plants.

Still other varieties are immune. A variety that is immune to virus X is immune to all the races of virus X that have been tested so far. The varieties immune to virus X harbor this immunity not only against contact and sap inoculation but also against infection in stalk and tuber grafts. Grafting a scion that is susceptible to virus X onto a stock that is immune to virus X results in formation of aerial tubers on the scion and provides a method for identifying varieties immune to virus X in segregating progenies.

Protective inoculation against races of virus X involves inoculating a susceptible variety with a weak race of the virus for protection against infection from stronger or more virulent races of the virus. As the stronger races of virus X cause greater reductions in yield than the weak races, the potato stocks inoculated with weak races are protected against this reduction.

Some varieties susceptible to virus A contract it less easily than others. Some varieties are immune. Some are symptomless carriers, manifest top and tuber necrosis, and are light green and rugose, or show mottling. Immune varieties are of two types virus-A, immune to aphid but not to graft infection, and virus-A, immune to both aphid and graft infection. Varieties included in the second type of virus A immunity manifest this immunity by top and tuber necrosis, a hypersensitive reaction. Examples of varieties of the second type of virus A immunity are Irish Cobbler, Earlaine, and segregating progenies of those varieties. Examples of the first type of virus A immunity are Katandin and Chippewa's well as segregating progenies of them. Observations on the reaction of different varieties to virus Y disclose that many of them develop dead spots on the leaves and streaking of veins, petioles, and stems, which is associated with brittleness and leaf drop. Rugosity and vein clearing appears on some varieties. Other varieties manifest light-green and rugose leaves to so slight a degree that they may be mistaken for healthy plants.

Although no varieties immune to virus Y have been found, some varieties are highly resistant in the field. Katandin and Chippewa are field resistant to virus Y under average conditions to aphid infestation. Under heavy aphid infestation, however, they contract considerable virus Y infection, an indication that the resistance is affected by the number of aphids.

In reaction to leaf roll, caused by the phloem necrosis virus, some varieties develop distinct foliage chlorosis, roll, and dwarfing. Others manifest those symptoms to a slight degree. Varieties vary in resistance to leaf roll. Under average field conditions Katandin contracts leaf roll in less than 10 per cent of the plants where less resistant varieties contract this disease in more than 50 per cent of the plants. Some of the new seedling Potatoes are more highly resistant to leaf roll than Katandin.

Investigations on potato diseases by the United States Department of Agriculture and State experiment stations have contributed much helpful information on identification, means of spread, disease resistance, and control of late blight, scab, ring rot, leaf roll, and viruses A, X, and Y. There is need for additional knowledge on the causal agents as to the origin and development of races; pathological anatomy; disease development in relation to moisture, temperature, soil reaction and nutrition; and the nature of disease resistance.

Problems in control involve devising and evaluating new fungicides and studies on antibiotics designed to find a product that is systemic and persists during the growing season. A primary objective in disease resistance involves breeding varieties that are immune from as many of the major potato diseases as possible and that possess other desirable characters such as quality, yield, and adaptability.

FUNGI CAUSES ROT IN SWEETPOTATOES

Diseases destroy 20 to 40 per cent of the sweetpotato crop in the field, in storage, and in transit to market. The average annual production from 1939 to 1948 of nearly 62 million bushels had a farm value of about 119 million dollars. The losses thus might be set at 24 million to 48 million dollars each year. More than 40 fungus diseases attack sweetpotatoes. Most of the damage is caused by seven of them. Diseases that affect the growing crop reduce the yield and the proportion of choice sise potatoes and cause disfigurations and skin blemishes that reduce their salability. Those that affect the potatoes in storage, in transit, and at the market cause decay, shriveling, and unattractive appearance.

The greatest monetary losses arise from damage in storage, in transit, and at the market. Sweetpotatoes are extremely perishable after they are dug and many decay before they reach the consumer. The later the decay occurs, the greater the monetary losses because of the added costs of harvesting, storing, grading, cleaning, packing, shipping, and marketing.

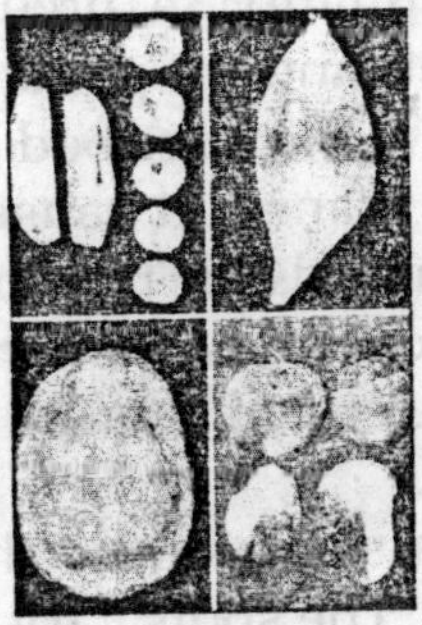

Upper left: Sections of sweetpotato roots affected with internal cork, caused by a virus. Upper right: Black rot of sweetpotato principally a storage rot. Lower left: Potato tuber infected with ring rot, a bacterial disease. Lower right: Fusarium tuber rot of Irish potato, a destructive disease in the United States.

Stem rot is one of the worst field diseases. It is found in all our growing areas. It destroys more than 50 per cent of the plants in some fields. It lowers total yield and the number of medium sise, uniformly shaped sweet-potatoes preferred for marketing.

Stem rot is most conspicuous about 2 weeks after the sprouts are set in the field. By that time many of the diseased sprouts are dead or are yellow and dying. The inside of the stem and vines of infected plants are brown or black instead of the normal white colour. The stem is often split near the ground level and decayed. Some of the diseased plants develop new roots above the decayed part of the stem and survive the entire growing season. The potatoes produced by the infected plants are usually small and decayed at the stem end. Their vascular tissues are brown. Sometimes a tuft of new sprouts grows from the stem end at harvest. Diseased sprouts growing from diseased seed potatoes in the plant bed sometimes can be recognised by a faint purplish tint, which shows through the white part of the stem and is caused by the dark tissues.

Stem rot is caused by Fusarium oxysporum f. batatas or F. hyperoxysporum. The fungi can live indefinitely in the soil. They also overwinter in stored sweetpotatoes that are diseased. They are usually spread from one locality to another by the exchange or sale of diseased seed potatoes or plants. They also may be carried in soil that adheres to farm implements.

Varieties of the Jersey group, including Big Stem Jersey, Little Stem Jersey, Maryland Golden, and Nancy Hall, are very susceptible to stem rot. Porto Rico is intermediate in resistance. Southern Queen, Triumph, and Yellow Strassburg are quite resistant but most of the more resistant varieties have poor table quality. A breeding programme has been started to develop resistant kinds of good eating quality.

Black rot is serious in the seedbed and field. It is especially destructive in storage and transit. It causes some reduction in yield, and some of the potatoes have visible black rot spots on them at harvest. Frequently, however, potatoes that appear free from black rot at harvest are so badly diseased at the end of storage or on arrival at market that they are nearly a total loss.

The most conspicuous symptom is the circular, dark-brown or black spots on the potatoes. The spots vary in sise from mere specks to 1 or 2 inches in diametre. The blackening usually extends into the potato only as far as the vascular ring, but in some varieties it extends much farther. Sometimes small, black bristles, the fruiting bodies of the fungus, develop at the centre of the spots. Black rot may cause small, round black spots on the white underground stem of the sprouts or it may cause :blackening and decay of most of the base of the stem, a condition that is sometimes referred to as black shank.

Black rot is caused by the fungus Ceratostomella fimbriata, which may remain alive in the soil for several years. It also lives from one season to the next on infected sweetpotatoes in storage. It is usually spread from one farm area to another by the exchange or purchase of plants and seed potatoes.

Harvested potatoes become infected by spread of the fungus spores during grading, brushing, and washing. When lots containing a few infected potatoes are washed, the wash water becomes contaminated by the fungus spores and nearly all of the potatoes become infected. Such potatoes may appear sound when they are packed for shipping after washing, but they may be heavily infected when they arrive at the market. The fungus grows best at temperatures of 73 to 81 F., but also makes good growth at 55 , the lowest temperature at which sweetpotatoes should be stored. Little or no growth Occurs at temperatures above 95 . When infected sweetpotatoes are held at 110 for a day, the fungus is killed. Most infection takes place through wounds.

No commercial variety of sweet-potatoes is entirely immune to black rot, but several unnamed seedling varieties have been discovered to be highly resistant. Soft rot is the most serious disease in storage and transit. It may start soon after the potatoes are stored and destroy nearly all of the potatoes before the end of the storage period or it may not affect the potatoes until they have been removed from storage and cleaned, packed, and shipped to market.

It is easily recognised by the soft, mushy decay and the accompanying fuzzy gray and black mold. The affected tissues are moist, brown, and stringy. The decaying potatoes have a rather pleasant aromatic odour, which is quite noticeable in storage houses when potatoes are affected with the disease. The affected potatoes lose moisture rapidly and finally become dry and brittle.

Soft rot is usually caused by the common bread mold fungus, Rhizopus nigricans, but several other species of Rhizopus also may cause it. Rhizopus spores are almost universally present, because the bread mold fungus grows on a wide variety of vegetable material.

Species of Rhizopus responsible for soft rot grow well at the usual storage temperatures and even better at higher temperatures. Infection rarely takes place except at wounds, such as those where the potato is broken from the stem and at cuts and bruises caused in digging, cleaning, and packing. Less infection occurs in bruised potatoes that are held at a high relative humidity and about 85 F. temperature for a few days immediately after wounding, because the injuries heal so rapidly at this temperature and high humidity that the fungus cannot get established.

All varieties are susceptible to soft rot, but some decay more rapidly than others. Southern Queen and Nancy Hall are quite resistant. Porto Rico, Big Stem Jersey, and Triumph are intermediate. Yellow Jersey is very susceptible.

The brown stain is only skin deep. It is usually worse at the stem end of the potato. It occurs as scattered blotches on lightly infected potatoes, but large areas may be brown or nearly black on badly infected ones. Scurf is caused by Monilochaetes infuscans, a fungus that overwinters on potatoes in storage. It persists in the soil for several years. If potatoes are bedded, the fungus grows from the mother potato up to the base of the sprouts. After the sprouts are planted it spreads down to the new potatoes. Most of the infection apparently comes from the seed potatoes and is carried to the field on the sprouts.

Scurf is worse on heavy soils and those containing a large quantity of organic matter than on sandy soils. Pox, or soil rot, is a widely distributed field disease, but generally it is less important than stem rot or black rot. It reduces the yield and the proportion of salable potatoes. Its most conspicuous symptom is the pits it causes on the potatoes. The pits are one-fourth inch to more than an inch in diametre and have a jagged margin. In the early stages the infected spots are dark-colored and water-soaked.

Later the skin of the potato covering the pit breaks and the contents fall out, leaving an empty cavity. The disease also kills the young feeding roots and causes dark lesions on the part of the stem below the soil line. Affected plants are usually stunted and have yellow leaves. Many of them die early.

Streptomyces ipomoea 'the causal fungus, lives indefinitely in the soil and is spread in soil adhering to farm machinery or plants, by wind-blown soil, and by floodwaters. It likes soils that are less acid than pH 5.2. The addition of sulfur to the soil to make it more acid has reduced damage from pox in some places. Internal cork, a new virus disease, was discovered in South Carolina in 1944. Since then it has been found in sweetpotatoes in many other areas, but it is most serious in South Carolina, Georgia, and North Carolina.

Sweetpotatoes affected with internal cork appear normal externally, but have dark brown to blackish corky spots scattered irregularly through the flesh. The corky tissues remain firm during cooking and have a slightly bitter taste. They vary in sise up to about one-tenth inch in diametre and one-fifth inch in length. Closely grouped corky spots may affect large areas in the flesh.

Leaves of plants affected with internal cork are sometimes marked with purplish ring spots. Only a little internal cork is found when the sweetpotatoes are dug but corky spots increase in number and sise during storage. The rate of increase is more rapid at 70 F. than at the recommended storage temperatures of 55 and 60 .

Java black rot ranks next to soft rot and black rot in importance as a storage disease. It occurs in all parts of the United States where sweet-potatoes grow and in many other countries. It was named Java black rot because it was first discovered on sweetpotatoes sent from Java. It is caused by a fungus, Diplodia tubericola. It causes a dry rot of the roots. The decayed tissues, brown at first, turn black and hard. The fungus forms black protuberances on the surface of the potato. The decay usually starts at the ends, but sometimes at breaks on other parts of the potato. It progresses slowly and there is little evidence of the disease until about a week after infection. The potato rots in 4 to 8 weeks. Surface rot, a common storage disease, causes shallow, circular, depressed spots on the surface and a gradual drying out.

It progresses so slowly that a great deal of damage may be done before the extent of the disease is recognised. The spots usually are not more than three-fourths inch in diametre and seldom penetrate below the vascular ring. They usually are grayish brown, but sometimes they may be so dark-colored that they resemble black rot.

The disease is caused by the fungus Fusarium oxysporum. Infection apparently takes place through the small rootlets that are damaged when the potatoes are harvested. The disease does not become conspicuous until the potatoes have been in storage for 2 months. Surface rot is worse on potatoes harvested when the ground is wet and when the potatoes are in the storage house several days before the curing process is started.

All varieties are affected, but Big Stem Jersey and Little Stem Jersey are more susceptible than some of the darker skinned varieties. The most effective and practical control of the field diseases is to plant only healthy seed potatoes and plants in healthy seedbeds and fields. Except for the fungi that cause stem rot and pox, none of the fungi that cause major diseases of sweetpotatoes remains alive in the soil for more than 2 or 3 years. A crop rotation of about 4 years generally rids a field of the fungi that cause black rot and scurf.

Plant beds should be thoroughly cleaned and filled with new soil or sand or new plant beds prepared. Plants free of black rot or scurf can be had even from affected sweetpotatoes by planting vine cuttings or sprout cuttings instead of pulled sprouts. The cuttings will be free of those diseases because black rot and scurf do not affect the above-ground parts. It is more practical to use sprout cuttings than vine cuttings because they can be obtained and Planted as early as pulled sprouts; vine cuttings cannot be made until much later.

The sprout cuttings should he made by cutting the sprouts about 1.5 inches above the soil line of the Plant bed without disturbing the below-ground parts. Once black rot and scurf have been eliminated from the sweetpotato seed stock, it is all right to use pulled sprouts in succeeding years. Because the fungi that cause stem rot and pox live indefinitely in the soil, neither the use of clean planting stock nor crop rotation will eliminate those diseases from a farm. The most practical control is by the use of resistant varieties when suitable ones have been developed. The addition of sulfur to the soil to lower the reaction to slightly less than pH 5.2 reduces the amount of pox. Much of the losses caused by black rot and scurf in storage and transit may be avoided by storing and shipping only potatoes that are free of those diseases. Black rot and scurf may be eliminated from the potato stocks by use of clean plants and crop rotation. Losses in storage and transit caused by soft rot, Java black rot, and surface rot can be greatly reduced by proper curing so that wounds caused in harvesting, handling, washing, and packing will heal rapidly and wall out the decay fungi. Temperature and humidity are important in curing sweetpotatoes. The temperature should be about 85 F. and the humidity should be high.

Only enough ventilation should be used to keep the potatoes from becoming wet. Frequently outside temperatures are high enough to favour healing when the sweet-potatoes are harvested and additional heat in the storage house is not necessary. Sweetpotatoes may be cured to reduce decay after they are taken from storage and washed and packed for shipment as well as when they are placed in storage in the fall. Storing sweetpotatoes at temperatures even slightly below 55 F. causes chilling injuries that make the sweetpotatoes more likely to decay.

PREVENTING THE DISEASES OF PEANUTS

Prevention, rather than cure, is the rule with the diseases of peanuts. Some beneficial preventive measures are the application of fungicides to seed to prevent seed rot, the use of fungicidal dusts or sprays on growing plants to prevent leaf spot, and the treatment of soil with chemicals before planting to control nematodes and soil insects.

Prevention involves more than the use of chemicals, however. Management practices that promote vigorous growth enable the plants to escape infection or to survive after infection. Outstanding are rotations that include one or two crops of corn, oats, or similar crops before each peanut crop; adequate mineral fertilisation; planting on light-textured, well-drained soils; and reduction of mechanical damage by cultivating implements.

Seed rot is the most serious of the diseases that affect stands of peanuts. It is caused by mildly pathogenic or saprophytic organisms that live in the soil mainly species of Fusarium, Rhizopus, Mucor, Diplodia, Penicillium, and Aspergillus. Their entrance is facilitated by broken seed coats and by other mechanical injuries produced by the sheller. Seed rot is most destructive under conditions that retard germination cool, damp weather, abnormally deep planting, and waterlogged soils. Most seed rot occurs within the first week after planting. Beginning about 1940, when farm labour became scarce, a trend was started towards the use of machine-shelled seed. Poor stands often resulted. Work started by Luther Shaw in North Carolina in 1939; and later confirmed by research in Virginia, South Carolina, Georgia, Florida, and Alabama showed that machine-shelled seeds treated with a seed protectant before planting produced stands comparable to those obtained by using hand-shelled and treated seed.

Although seed usually are treated at the time they are shelled because of convenience, the treatment may be applied with equal effectiveness just before planting. Several good seed protectants are available commercially. Eight materials were tested in Alabama in 6 years. Best results were had with Ceresan M (7.7 per cent ethyl mercury p-toluene sulfonanilide). Phygon (90 per cent 2, 3-dichloro-1,4-naphthoquinone), Arasan (50 per cent tetramethyl thiuramdisulfide), and Spergon. (98 per cent tetrachloro-p-benzoquinone) were somewhat less effective. Reasonably good results were obtained with Yellow Cuprocide (yellow cuprous oxide containing 47 per cent metallic copper) and Dow 9 B (50 per cent zinc trichlorophenate). Seedox (50 per cent 2,4,5- trichlorophenyl acetate) and Merc-O- Dust (a compound of indefinite composition containing mercury and formaldehyde) were ineffective. Similar results have been obtained in other States.

Although the mercurial treatments such as Ceresan M are most effective, they are more dangerous to use. Over-dosage results in poor germination. The radicle of the germinating seed does not elongate but becomes thickened and stubby. If a seedling is produced, it is stunted and never makes normal growth. Therefore and because of the danger to warm-blooded animals that might eat treated seed most States recommend one of the organic materials such as Arasan or Spergon in preference to the mercurial treatments.

The organic seed treatments have a tendency to be somewhat erratic in performance. In tests lasting 6 years at Auburn, Ala., Ceresan M was the most consistent of four treatments. Spergon was the best of the four in 1949 but the poorest in 1947. Dow 9 B was practically as good as Arasan or Ceresan M in 1946 but was considerably poorer than either in 1948. Arasan gave less protection than Ceresan M in 1944, 1946, 1948, and 1949.

Seed protectants are applied as dust treatments at the rate of 2 or 3 ounces to 100 pounds of seed. Slurry treatments liquid suspensions of the fungicide have not been popular on peanuts, probably because of the tendency of the seed coats to peel after the slurry treatments are applied. The tendency appears to be more common with Spanish-type seed than with seed of runner peanuts.

Seedling blights may be destructive in some localities, but generally they do little damage in established stands. Damping-off is not a serious disease of peanuts. One of the most common types of seedling blight in peanuts is caused by Sclerotium bataticola. This organism causes a disease known as charcoal rot in a number of other species of plants during periods of high temperatures. Infections result in the formation of lesions on the succulent stems.

The initial infections usually are near the soil line, but the lesions may extend downward some distance into the soil. If plant growth is retarded by dry weather and if the temperature is high above 75 F. the lesion is likely to girdle the stem and kill the plant. The stem assumes a dull brown Colour and becomes quite dry. Sometimes the progress of the disease is stopped and the seedling is stunted rather than killed. Then the plant is made more susceptible to other diseases later in the season. Charcoal rot in peanuts can be recognised by the many small, irregular, black sclerotic that develop in irregular affected tissues and give them a dark gray or black colour on the inside.

Another form of seedling blight is dry rot, caused by Rhizoctonia solani. It is like charcoal rot in the early stages. Lesions develop on the stem near the soil line. The plant may be girdled and killed; if the infection does not spread, the plant may partly recover. The lesions are not likely to extend as far below the surface as those caused by Sclerotium bataticola.

There are no specific control measures for seedling blights. Losses may be reduced by planting treated seed of good quality on a well-prepared seedbed. Of the diseases of growing plants, leaf spot generally is the most destructive disease during the growing season. It is recognised by the brown or black and somewhat circular spots on the leaves. As the disease progresses, the spots enlarge until the entire leaf is affected. Defoliation follows. The detrimental effects of leaf spot are threefold: The yield of nuts is reduced; the quality of the peanut hay is lowered, and the fallen leaves provide organic matter on which inoculum of other fungi, such as Sclerotium rolfsii, is produced.

Two species of fungi cause leaf spot. Each produces characteristic symptoms. Early leaf spot, caused by Cercospora arachidicola, produces spots that are light tan at first. With age the spots become reddish brown to black on the lower surface and light brown on the upper surface of the leaf. A yellow halo surrounds each spot. Late leaf spot, caused by C. personate, produces dark-brown or black spots on both surfaces of the leaf. The spots usually are somewhat smaller than those of early leaf spot and there is no distinct halo. The fungi can be distinguished by microscopic examination of the conidia.

The conidia of Cercospora arachidicola are colorless to pale olive green and often curved. According to measurements made by W. A. Jenkins in 1938, they are 37-108 by 2.7-5-4 microns and contain 3 to 12 septations. The conidia of C. personata are much shorter and considerably thicker; they measure 18-60 by 5-11 microns, with 1 to 8 septations. They are generally cylindrical and seldom curved. Both leaf spots attack the leaves, petioles, pegs, and pods of Spanish, bunch, and runner peanuts. No host plants other than Arachis hypogaea are known for them.

Primary infections of leaf spot are caused by ascospores formed in the spring on overwintered peanut leaves. Secondary infections result from conidia. Although the fungus has been reported to be seed-borne, this method of overwintering is of minor importance. The principal means of dissemination is by wind-borne inoculum. The ascospores or conidia germinate within a few hours and penetrate the leaf directly through the epidermal cells or through stomata. In both species, the mycelium is intercellular at first. Branched haustoria are soon formed by the mycelium of Cercospora personata and the host cells are not killed outright. C. arachidicola does not form haustoria; the host cells are killed in advance and the invading germ tubes enter dead cells.

Infections on leaves are visible within 8 to 23 days after inoculation. Leaf spot is controlled by fungicides applied as dusts or sprays to the foliage. Most commonly used are dusting sulfur or dusting sulfur containing approximately 3.5 per cent metallic copper. Specific recommendations vary slightly in different States; three to five applications beginning about 90 days after planting often are recommended.

Usually applications are repeated every 10 to 14 days; 15 to 25 pounds of dust per acre are applied each time. Bordezaux spray-6 pounds of copper sulfate and 2 pounds of hydrated lime in 100 gallons of water is effective, but offers no particular advantages over fungicidal dusts. The dithiocarbamate fungicides have not come into general use on peanuts, probably because of their higher cost. Increases in yield resulting from dusting depend upon the fertility of the soil on which the crop is grown, the prevalence and severity of leaf spot, and the type of peanuts being grown.

Highest returns have been obtained on runner peanuts growing on fertile sandy loam soils in the Southeast. Increases of 1,000 pounds an acre have been reported by the Alabama Agricultural Experiment Station. The increases generally amount to 500 pounds an acre or less. Southern blight, caused by Sclerotium rolfsii, is more destructive on

Spanish and bunch types of peanuts than on runners. It may appear on the plants at any time during the growing season but is more likely to cause damage in late summer and early fall as the plants approach maturity. The fungus, soil-borne, attacks the plants near the soil line. It causes wilting and eventual death of the part of the plant above the infection.

It may attack the central stem and affect the entire plant, or the infection may be limited to one or more branches. Diseased plants wilt; the leaves gradually turn brown or black and eventually fall off. The fungus destroys the succulent tissues in the stem; the vascular bundles that are left give a shredded appearance to the diseased area. When the humidity is high, light-tan to reddish-brown sclerotia are formed in abundance on the infected tissues and the surrounding ground.

The sclerotia are usually spherical and about the sise of mustard seed but may be larger and irregular in shape. In dry weather the sclerotia are not so conspicuous. Sclerotium rolfsii attacks hundreds of species of plants. In the absence of suitable hosts it lives quite well as a saprophyte in the soil. Crop rotation is therefore of limited value for control. Most grasses are highly resistant, but among them usually are enough volunteer plants that are susceptible to enable the fungus to survive in the soil. Southern blight appears to be more destructive on peanuts following peanuts or cotton, however, than on peanuts following corn.

Collar rot of runner peanuts is a disease complex that sometimes resembles southern blight. The disease has been reported only from Georgia, Florida, and Alabama. The cause has not been definitely established. The fungi that can be isolated from diseased plants are either saprophytic or weakly pathogenic, and efforts to produce the disease by inoculating healthy peanut plants with these fungi usually are unsuccessful. Diplodia theobromae apparently is always associated with the disease. Other fungi that usually are present include various species of Fusarium, Penicillium, Trichoderma, and Rhizopus nigricans. Sclerotium rolfsii and S. bataticola sometimes are isolated.

Collar rot may result when plants that have been wounded are invaded by any of those fungi. L. W. Boyle in Georgia has demonstrated that young seedlings of peanuts are subject to sunscald. The injured areas on the stem or branches become lesions or cankers through which the fungi gain entrance. If the lesion is small, the plant may continue to grow for several weeks before the disease becomes severe enough to cause death. Other types of injury also may pre-dispose the plants to attack injury by cultivating implements, windblown sand, and chewing insects. It is Possible that infections also develop in lesions produced by the charcoal rot and rhizoctonia dry rot organisms. Collar rot may appear at any time during the growing season. It usually is most noticeable during July, August, and early September. It may involve the entire plant or only a single branch. The affected part wilts suddenly; the leaves become pale green and turn dark brown or black rather quickly.

Death is usually complete within 2 or 3 days after wilting becomes evident. If the entire plant is diseased, the wind usually blows the leaves and stems away, leaving the ground completely bare. The taproot is usually shredded at or near the soil line. *There is no control for collar rot*: Nematodes have become increasingly important on peanuts. Several species attack peanuts, among them the root knot nematodes (Meloidogyne arenaria and M. hapla), meadow nematode (Pratylenchus leiocephalus), and the sting nematode (Belonolaimus gracilis).

The root knot nematode attacks both roots and fruits and causes the formation of galls. The feeding roots are deformed. The knots on the roots usually are similar in sise and shape to the nodules formed by nitrogen-fixing bacteria. On the fruits and peduncles the galls are corky and variable in shape. Symptoms on the above-ground part of the plant are not characteristic. Plants are stunted; leaves are likely to show chronic symptoms of wilting; and the colour is light green. In advanced stages leaves are necrotic along the margins, and considerable leaf shedding results. Heavy infestations may reduce yields to 500 pounds or less to the acre.

The symptoms of meadow nematode injury consist of lesions on peanut shells and lateral feeding roots. The damage it does to plant tissues provides avenues of entrance for soil fungi that cause peg rot. The most characteristic symptom of the sting nematode is an unusually long taproot with relatively few lateral roots. Those that are present are short, stubby, and deformed. This nematode causes damage by puncturing the roots with a needlelike "sting" and withdrawing the juices. The nematode does not enter the roots but feeds externally. The area punctured by the sting is attacked by soil fungi. Small dead spots develop; the result is root pruning, which accounts for the elongated taproot and the deformed lateral roots.

If the nematode infestation is light, control can usually be accomplished by rotation. Such crops as corn, oats, and grain sorghum are resistant to the root knot nematode but not to the sting nematode. Control by rotation is not successful if infestations are heavy. If infestations are heavy, nematodes can be controlled by fumigating the soil before planting. A fumigant, such as Dowfume W-40, should be applied to the drill at the rate of 7 gallons to the acre about 3 weeks before planting. After fumigation the soil should be left undisturbed until planting time. Care should be taken to plant the seed directly above the area in which the fumigant was applied. The plows used for opening the furrows should be as small as possible to prevent mixing of fumigated and unfumigated soil.

The root knot nematode, Meloidogyne arenaria, attacks several plants other than peanuts, including sweet-potato, Ramie, coffeeweed, tomato, okra, onion, peach, and several weeds. Peg rot, strictly speaking, is not a disease of the growing peanut plant. The term includes all losses that result from rotting, sprouting, and other types of damage to the developing nuts or to loss of nuts in the soil at harvest by breaking of the gynophores.

The peanut flower is formed in the axil of the leaf on the stem. After pollination, the petals wither, and the peg (or gynophore) elongates and grows into the soil, where the fertilised ovary at the tip of the peg enlarges to form the pod or peanut fruit. The peanut fruit thus develops in an environment populated with soil micro-organisms and is subject to attack by them during all of its development.

Soil insects such as wireworms and the larvae of the southern corn root-worm do considerable damage to the developing pods and facilitate infections by fungi. Many soil fungi attack the slender gynophore near the soil line as the fruit approaches maturity and many gain entrance to the tissues. Their action weakens the gynophore so that when the plant is pulled or lifted from the soil the gynophore breaks, leaving the pod in the soil. Early defoliation of plants by leaf spot, leaf-eating worms, or other causes also causes the gynophores to be more brittle and to break more easily.

Injuries to the "shell" of the developing fruit by soil insects permit soil-borne fungi to gain entrance. Discoloration or decay of the seed follows. If harvest is delayed beyond maturity of Spanish or bunch peanuts, the seeds begin to germinate. Miscellaneous fungi, including Sclerotium rolfsii and S. bataticola and species of Rhizoctonia, Diplodia, Rhizopus, Aspergillus, and Penicillium, may be isolated from developing pegs and mature fruits. Apparently any one of them may cause one form of peg rot under proper conditions.

Several other diseases of growing peanuts occur occasionally in the United States. Peanut rust, Puccinia arachidis, has been found occasionally in Florida, Alabama, and Texas. A leaf spot caused by an unidentified species of Phyllosticta appears on young peanut leaves early in the spring, but infections do not spread to older leaves. Botrytis late blight of peanuts has been observed on over-mature peanuts during cool, damp days of early fall in Georgia. It is sometimes destructive on peanuts grown in greenhouses in winter. Fusarium wilt of peanuts has been reported from Southeastern States many times, but apparently it has not caused heavy damage. Virus diseases and bacterial wilt destructive in the East Indies and in Africa are not serious in the United States.

As the peanut develops in the soil, a variety of fungi become established in the shell, the peg, and occasionally the seed. When peanuts are harvested, they usually are placed in stacks or small piles or windrows and allowed to cure. During the curing process the fungi continue their activity and frequently become established in the tissues of the seed or on the interfaces of the cotyledons. Activity of the fungi during curing, together with the activity of the enzymes within the peanut seed, results in various types of seed deterioration. In commercial circles all these troubles are grouped together under one heading damage.

Concealed damage is first evident as a slight yellow discoloration on the interfaces of the cotyledons. A mat of fungal mycelium develops between the cotyledons as the disease progresses. The tissues become rancid and discolored. There is no evidence Of the disease on the outside until its late stages. If conditions remain favourable, the entire seed becomes shriveled and blackened, assumes an oily appearance, and is unfit for human consumption. The per centage of free fatty acids in the oil increases markedly. The disease is more common in runner peanuts than in Spanish or bunch peanuts, and has been more prevalent in the southeastern peanut belt than in the Virginia-North Carolina area and the Texas-Oklahoma area.

Seeds with concealed damage may contain a variety of fungi, primarily Diplodia theobromae. Others are species of Fusarium, Aspergillus, Penicillium, Rhizopus, and Rhizoctonia. Occasionally, Sclerotium ro4fsii and S. bataticola are found. The disease has been produced by inoculation only with Diplodia theobromae and Sclerotium bataticola. Although initial infections usually occur before digging, concealed damage makes its most rapid development during the curing process.

The moisture content of the seed is the principal conditioning factor, as temperatures during the curing period are usually near the optimum for fungal development. Concealed damage develops most rapidly in partly cured peanuts that contain 15 to 35 per cent moisture. The disease does not develop appreciably in seeds containing less than 10 per cent moisture. Neither soil type nor fertilizer practice has any measurable effect on development of the disease. Concealed damage can be controlled by rapid curing of peanuts as soon as they are harvested and the use of resistant varieties. Freshly dug runner peanuts usually contain 40 to 55 per cent moisture. If peanuts are allowed to wilt before being placed in symmetrical, well-capped stacks with provisions for air circulation at the base, curing usually is rapid enough to prevent development of concealed damage.

If peanuts are stacked before wilting or if they are allowed to become brittle before stacking, curing is retarded and the disease is more pronounced. Peanuts that are cured in windrows instead of stacks usually contain very little concealed damage, but this method of curing is more hazardous because of shattering in unfavourable weather. The Dixie Runner variety has a great deal of resistance to concealed damage, and that resistance has been incorporated into a number of other strains of peanuts. By using resistant varieties and by exercising reasonable care at harvest, most growers can keep losses from concealed damage to a minimum.Blue damage is a discoloration that often occurs on the seed coat of Spanish peanuts cured during periods of warm, damp weather. The small spots usually have a "bull's-eye" appearance, with colours ranging from blue through black. Larger spots may be irregular in outline with no evident centre. The discoloration does not produce detectable changes in chemical constituents, taste, or germination of seed. Quite often the discoloration does not extend into the cotyledons, but sometimes the cotyledons may be conspicuously discolored.

Experiments at the Georgia Agricultural Experiment Station showed that the discoloration results from an interaction between pigments in the seed coat and oxalic acid produced by Sclerotium rolfsii. Under favourable conditions S. rolfsii grows saprophytically on the peanut shells during the curing process. The fungus secretes oxalic acid, which diffuses into the tissues in advance of the hyphae. When curing conditions are unfavourable for growth of S. rolfsii, no discoloration occurs.Other storage disorders, including seed rot, rancidity, and reduced germination, may be the result of improper curing or of improper storage. Peanuts should contain less than 10 per cent moisture at the time they are stored. Some deterioration occurs at moisture contents between 6 and 10 per cent, but the rate is fairly slow. The relative humidity of the storage environment should be 65 per cent or less. Under those conditions, well-cured peanuts will keep well in storage for several months.

WAYS TO COMBAT DISORDERS OF TOMATOES

Every tomato grower is faced with the hazard of loss from disease. Infections by fungi, bacteria, or viruses are responsible for much of the loss, but extremes of temperature and moisture or excess or deficiency of mineral elements in the soil at times may damage the crop. The symptoms thus caused can be classified as wilts, leaf spots, fruit spots or rots, and abnormalities of growth of the foliage or fruits.

Fusarium wilt, caused by the fungus Fusarium oxysporum f. lycopersici, is serious on tomatoes. Its first symptom is a yellowing of the oldest leaves, usually on a single stem. Beginning at the base, the yellowed leaves wilt and die until all the leaves are killed. Often a single shoot dies before others are severely injured, but eventually the entire plant is affected. There is no soft decay of the stem, but if one cuts it lengthwise near the ground, he can see in the woody portion next to the outer "bark," or cortex, a brown discoloration of the water-conducting tissues; fibrovascular bundles which is characteristic of the disease.

The fungus that causes fusarium wilt can persist in the soil for many years. Infection occurs through the roots. From them the fungus passes into the water-conducting vessels of the stem. Apparently it produces a toxic substance that causes the wilting of the foliage and eventual death of the plant. The fungus is most active at temperatures between 80 and 90 F. Losses from the wilt are most common and severe in the warmer sections of the country. Clean fields can be infested by planting diseased seedlings, or by drainage water, farm implements, or any agency that carries infested soil even in small amounts to the clean field.

Other wilt diseases of tomato are caused by soil-borne bacteria and fungi. Unlike fusarium wilt, they are characterised by a decay of the stem near the ground and a wilting of the entire plant with little or no yellowing of the leaves.

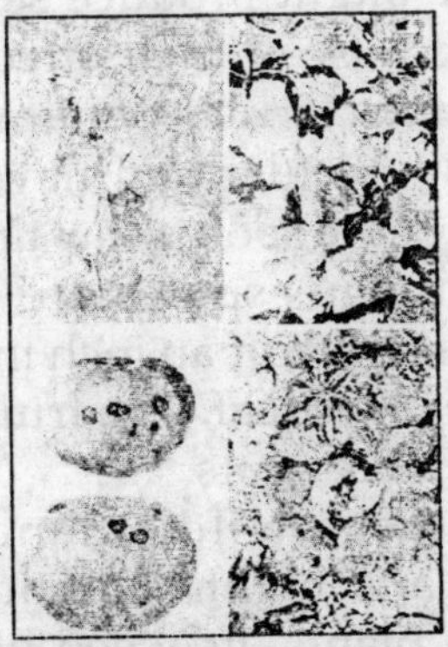

Upper left: Pea streak, caused by a virus, on pea pods. Upper right: Anthracnose of cucumber Plants; the disease also affects watermelon and muskmelon growing in humid regions. Lower left: Bacterial spot of tomato on two fruits. Lower right: A plant of Great Lakes lettuce affected with rib blight, a physiologic disease. Bacterial wilt, caused by Pseudomonas solanacearum, is an example.

That organism causes a watery decay of the pith, which soon causes the wilting and death of the plant. These wilt diseases are most common in the southern United States but are of much less economic importance than fusarium wilt. Leaf spot diseases of tomato cannot be described solely on a basis of leaf symptoms, since certain of their causal agencies also are responsible for serious spotting of the stems and spotting or rot of the fruits.

The most common and destructive leaf spots are caused by bacteria and fungi that produce spots in which the cells of the leaf have been killed. In some of the diseases an area of yellowed tissue surrounds the dead spots. With the exception of one fungus that destroys areas of tissue almost too large to be called spots, the organisms invade only a limited area around the point of infection and produce spots of characteristic sise and appearance.

They may be irregular or circular and vary from one thirty-second inch to one-half inch in diametre. Both the petioles and leaflets may be spotted, and those injuries commonly cause the death of many leaves. Some viruses also cause dead spots or patches on the leaves but the spots are of a type that usually is not confused with those caused by bacteria or fungi.

The leaf spot caused by the fungus Septoria lycopersici is an example of a disease that severely injures the leaves but does no appreciable harm to the fruit. On the other hand the fungus (Alternaria solani) that causes alternaria leaf spot and the fungus (Phytophthora infestans) that causes late blight produce severe injury to the foliage and are also important because of their effect on the fruits.

Septoria leaf spot is characterised by numerous water-soaked spots that soon become roughly circular and have gray centres with dark margins. Later the centres of the spots show tiny dark specks (pycnidia), the bodies in which the fungus spores are produced. The spots are from one-sixteenth to one-eighth of an inch in diametre and may occur in such numbers as to cover the leaf. The fruits rarely are affected, but stems and blossoms may have spots.

The fungus also attacks wild solanaceous plants, such as horsenettle, jimsonweed, and nightshade. It over-winters in the soil on the remains of tomato plants, its weed hosts, and other decaying vegetation. Spores produced on such material cause the infection of tomato leaves near the soil. Spores then are produced in the spots on those infected leaves and, in wet weather, are exuded on the surface of the spot. From there they are splashed to other leaves by rain and may be carried on the hands or clothing of farm workers or on farm equipment. The disease may occur on plants of any age, but in the field it seems to develop more rapidly after some fruit is set. If temperatures are moderately high and rains frequent, there is a progressive loss of foliage until only a few leaves may remain at the tips of the stems. When that occurs, yields are reduced and many of the fruits are injured by exposure to the sun. In hot, dry seasons the disease is much less damaging.

Septoria leaf spot causes serious loss in some sections of the Central and Atlantic States as far south as Arkansas, Tennessee, and South Carolina, but is of minor importance in the Mountain and Pacific States. The first symptoms of alternaria leaf spot (early blight) consist of small, dark-brown spots on the older leaves. The spots may enlarge until they are one-half inch in diametre and have an irregular or circular outline. The large spots often are marked with rings in a target pattern. The tissues surrounding the spots often turn yellow and this discoloration may cover the leaflet. The disease is common on seedlings, where the spots usually are small and dark. In the field, the spotting and yellowing of the leaflets, together with spotting of the petioles, often kills much of the foliage.

Plants in the field also show many spots on the stems. They are light brown, with darker margins, and from one-eighth to one-half inch in sise. On seedlings the fungus causes large, dark cankers (collar rot) that may girdle the stems. The varieties Southland and Manahill have high resistance to collar rot and some resistance to alternaria leaf spot. Alternaria solani also causes a disease of potatoes known as early blight. Occasionally it affects peppers.

It can live on decaying plant tissue in the soil. Most of the primary infection in tomatoes comes from that source, although contaminated seed may possibly cause occasional infection. Alternaria spot is the most common leaf spot disease of tomatoes in the Central and Atlantic States and occurs in all tomato-growing regions, although it is of less importance in the Mountain and Pacific States. Late blight is characterised by irregular, greenish-black patches, which first appear on the older leaves.

They rapidly enlarge until they may cover one-fourth to one-half of the leaflet. The spots soon turn brown and many of the infected leaves wither, yet frequently remain attached to the stem. When weather conditions favour a rapid development of the disease, so much of the foliage may be affected that the plants look as though injured.

by frost. The fungus also produces water-soaked, brown streaks on the stem. In moist weather there often is a white, downy growth of the fungus on the under surfaces of the spots on the leaves and stems. The fungus that causes late blight also causes a disease of the same name on potatoes. It is not carried in tomato seed and does not live over in the soil. It infects potato tubers, which can produce diseased plants in the spring. From them spores can be carried by the wind and infect plants in other fields.

Late blight of tomatoes is caused by a strain of the fungus somewhat different from that which ordinarily occurs on potatoes. The tomato strain can infect potatoes, however, and has been recovered from blight-infected potato tubers as a typical tomato strain. What appears to be the typical potato strain of the fungus may produce only mild symptoms when first transferred to tomatoes but can develop the virulence of a typical tomato strain after a number of passages through tomato plants.

Late blight commonly occurs on potatoes and tomatoes grown in the Gulf States during the winter and early spring. The disease works northward with those crops as the season progresses. Most of the canning tomatoes of the Central and Atlantic States are produced from seedlings grown in the South, and late blight occasionally appears in some plant fields.

If the infected transplants are shipped north they become a possible source of infection to other fields. Besides that possible hazard and that of infection from diseased potato plants, there also is an occasional carry-over of the fungus on greenhouse tomatoes in winter. Seedlings grown in those houses in the spring often become infected and carry the disease into the field. Cool nights and only moderately warm days, with abundant moisture, favour the rapid development and spread of late blight.

Late blight of tomatoes was of little or no importance in many tomato-growing sections until 1945, although the disease had caused some loss in the Northeastern, Middle Atlantic, South Central, and South Atlantic States. Occasional damage also had occurred in Texas and on the Pacific coast. In most of the sections where the processing tomatoes are grown east of the Mississippi, however, the disease was almost unknown.

In 1945 there were some losses in the North Central States. Early in the season of 1946 several unusually long periods of cool, moist weather were followed by an epidemic of the disease along the Atlantic seaboard, which destroyed nearly half the crop. The damage was all the worse because the growers had no previous experience with the disease and were not equipped to apply the fungicides that would have prevented much of the loss. Since then, late blight has occurred to some extent each year in many tomato areas of the Central and Atlantic States, but the loss has been greatly reduced by increased use of fungicides.

A number of fruit rots are caused by bacteria and fungi that produce no other significant injuries to the tomato, plant. The most serious are due to soil-borne fungi that can infect Uninjured fruits. There also are organisms that can only enter the fruit when the epidermis is broken by growth cracks, mechanical injuries, or the lesions caused by the common fruit-spotting organisms. Such decays are unimportant in the field but may cause losses of fruits in transit, in ripening rooms, and in the markets.

Anthracnose of tomato, caused by the fungus Colletotrichum phomoides, is the most serious of the fruit rots caused by organisms that produce little or no injury to the leaves. Uninjured fruits can be infected at any stage, but in green fruits the fungus remains latent and no spotting occurs until the fruits ripen. Ripe fruits show dark, depressed spots whose centres have concentric markings. The centres are tan with dark specks (acervuli). In moist weather the masses of spores, which give the spots a salmon-pink colour, are produced in the dark specks. The spots are circular and about three-eighths inch in diametre. They penetrate a short distance into the fruit and must be removed when the fruit is processed.

Anthracnose is a serious pest of processing tomato crops in the Central and Atlantic States, but is of little importance where tomatoes are picked green for shipment to distant markets. The fungus occurs in the soil and the first infections come from that source. The spores produced on the surfaces of the fruit spots later are splashed by rain or otherwise spread to healthy fruits. In warm, moist weather many fruits are thus infected.

Fruit spots also are produced by some organisms that spot the leaves. Some of them penetrate the epidermis of the fruit and produce spots that do not extend very deeply into the flesh. The spots vary from one-sixteenth to one-half inch in diametre and have a roughened, broken surface that is brown or gray in colour.

Alternaria leaf spot and late blight, however, are characterised by rots that cover a considerable area of the fruit and extend deeply into the tissues. Spotting and ring markings of the fruits also are associated with some viruses that cause spotting of the leaves. Losses from them are not of major significance. The fungus that causes alternaria leaf spot attacks the green fruits at the point of attachment to the stem.

Here the fungus produces dark, leathery, sunken spots, which have concentric markings and may be an inch or more in diametre. A dark, dry, decay extends well in the fruit and makes it unfit for use. This fruit rot phase of the disease, however, causes less damage than the defoliation caused by the leaf spotting.

The fruit rot associated with late blight often is a cause of major loss. Fruit infection occurs at any stage of growth and is most frequent on the upper half of the surface. The grayish green, water-soaked spots enlarge until they may cover half the fruit and eventually become brown, with a firm, slightly wrinkled surface. In moist weather, a downy, white growth of the fungus often appears on the surfaces of the spots.

Tomatoes are subject to a number of diseases caused by viruses that produce a mottling or yellowing of the leaves and are accompanied by varying amounts of curling, twisting, and malformation of the leaflets. Some of them also cause a streaking of the stem and a spotting of the leaves and fruits. Most of the viruses may occur wherever tomatoes are grown in the field or greenhouse, but the destructive virus known as curly top is confined largely to the Mountain and Pacific States as it is transmitted only by a leafhopper whose breeding grounds are limited to semiarid sections in the West.

Tomato mosaic, as the name is commonly used, is the most common virus disease of tomatoes. The name refers to the disease caused by the tobacco mosaic virus (Marmor tabaci), which also affects peppers. The common green strain of the virus causes a light- and dark-green mottling of the leaves, a slight curling, and occasional malformations of the leaflets. Infected seedlings are stunted, but plants that become infected after they reach the fruiting stage show no severe check in growth although yields are reduced. There also are "yellow" strains of the virus which cause a pronounced yellow mottling of the leaves and some mottling and blotching of the stems and fruits.

The tobacco mosaic virus is rarely, if ever, transmitted in commercial tomato seed. It does not persist long in field soils, although in greenhouses, where one crop may follow another in a few weeks, some infection may occur from plant remains in the soil. Perennial weeds, such as horse-nettle, groundcherry, and plantain, sometimes are infected with the virus and may be a source of infection. The virus remains active for years in dried tobacco leaves and is known to occur in pipe, cigar, and cigarette tobaccos.

Smokers, therefore, may carry the virus on their hands, and some of the initial infection of tomato seedlings may come in that way. The virus is transmitted so readily by successive contact with mosaic-infested and healthy plants that it is almost certain to become widespread after it appears in the greenhouse or field. Aside from its own effect on the plant, the tobacco mosaic virus is a menace to the crop because infected plants also may become infected with the virus X of potatoes (Marmor dubium). The virus is present in the tops and tubers of all the older varieties of potatoes but usually causes little visible evidence of disease.

Tomatoes that are infected with both the tobacco and potato viruses develop a disease known as double-virus streak. Such plants show small, dead, brown spots on the leaves, which are mottled, dwarfed, and curled. Stems and leaf petioles develop the narrow brown streaks that give the disease its name. The plants set few fruits and those produced are often misshapen and have irregular brown spots on their surfaces. Streak can be very destructive in the greenhouse where the necessity of handling the plants in pruning and tying leads to its rapid spread. Tomatoes also are sometimes infected with the cucumber mosaic virus (Marmor cucumeris).

Such infection is much less common than that with the tobacco mosaic virus. The leaves show only a mild mottling but often are dwarfed or malformed. Frequently the leaflets are reduced to little more than a central rib, and this "shoestring" symptom is one characteristic of the disease. The plants that are infected with the cucumber mosaic virus are dwarfed and often have an abnormally compact, bushy appearance. The few fruits they set are small and often malformed.

Unlike the virus of tobacco mosaic, the cucumber mosaic virus does not withstand drying. It can be transmitted by mechanical means, but on tomatoes that occurs less readily. It is commonly disseminated by aphids, particularly the species that feed on cucumbers and melons. As they prefer to feed on plants other than tomatoes, the aphids cause less infection on tomatoes than on certain other crops.

The virus is not transmitted in the seed and does not persist in the soil. It does infect a wide range of cultivated and wild plants and is common on cucumbers, muskmelons, celery, peppers, and some ornamental plants. It lives over winter in some ornamentals, such as perennial phlox, and in some perennial weeds, such as milkweed, groundcherry, and catnip. Aphids feeding on such infected plants in the spring later carry the virus to cultivated crops.

Some evidence of such diseases can be found in many fields every season. Usually the losses are minor, although the drop of blossoms, caused by hot, drying winds, sometimes is a serious factor. Blossom-end rot, a common non-parasitic disease, usually appears on tomato fruits when they are about half-grown. It consists of a brown discoloration that enlarges until it covers one-third or more of the fruit. The dark tissues shrink and form a flattened or slightly concave area with a dry, leathery surface. There is no rot of the fruit unless the dead tissue is later invaded by fungi or bacteria. Blossom-end rot usually appears when plants that have made a rapid growth are subject to high temperatures and drought. Then the cells at the blossom end of the fruits apparently fail to receive enough moisture to maintain their growth, and the tissues break down.

Plants that have had heavy applications of nitrogenous fertilizer are especially susceptible to blossom-end rot. Marked fluctuations in soil moisture increase the likelihood of injury. Plants that arc pruned and staked consistently show more blossom-end rot than those growing naturally in otherwise similar conditions. That may be due largely to the restriction of the root system that results from drastic pruning of the tops.

Disease control is essential in the profitable production of tomatoes. The first step consists of measures that help to prevent the occurrence of seedling diseases. Wherever possible, seedlings should be started in soil that has not grown diseased plants. If that is not feasible, the soil can be disinfected by treatment with formaldehyde or chloropicrin or by steam. Seed treatments are used to disinfect the surface of the seed and to coat the seed with a fungicide that will protect the seed and germinating seedling from fungi in the soil that cause seed decay and damping-off of seedlings. Tomato seed can be dusted with such materials as New Improved Ceresan, Semesan, Phygon, Arasan, or Thiram 50. They generally are used at a rate of 0.3 to 0.5 per cent of weight of seed.

Losses from wilt diseases and fruit rots caused by soil-borne organisms can be reduced somewhat by crop rotation and the avoidance of fields where such diseases have damaged previous tomato crops. Seedbeds should always be located on clean soil. If plants are purchased, they should be obtained from plantings free from disease. In the greenhouse, wilt diseases and injury from the root knot nematode can be controlled by disinfection of the soil by chemicals or steam. In the field, the use of steam is impracticable and chemical treatments, other than those effective only against nematodes, have as yet been too expensive for large-scale use.

Resistant varieties offer the most effective means of preventing losses from wilt diseases, but we do not yet have varieties resistant to all diseases of this type. Breeding of varieties resistant to fusarium wilt by the Department of Agriculture and State experiment stations has made it possible to maintain a profitable tomato industry in regions where the disease is prevalent. Some of the resistant varieties that have had most general use are Rutgers, Marglobe, Indiana Baltimore, Pearson, and Pritchard. They are not immune to wilt and occasionally are damaged by it. Since 1940, however, varieties have become available that have extremely high resistance. Among them are Pan America, Southland, Homestead, Cal 255, Manahill, Manasota, Fortune, Boone, Tipton, Kokomo, Tucker, Sunray, and Golden Sphere.

Another such variety, Jefferson, has been introduced by a private seed company. Resistant varieties suited to greenhouse use include Ohio W-R Globe (which possesses high wilt resistance), Blair Forcing, and Michigan State Forcing. A race of the fusarium wilt fungus, which can attack plants of the highly resistant varieties, has been reported in Ohio and Missouri, but we have had no report of widespread infection of those varieties in commercial plantings.

Varieties resistant to verticillium wilt, caused by the fungus Verticillium albo-atrum, have been developed for use in sections in the West where the the disease has caused serious losses. The varieties Riverside, Essar, and Simi have tolerance of verticillium wilt but are injured when conditions are very favourable to the development of the disease. They also have moderate resistance to fusarium wilt. Two newly introduced varieties, Loran Blood and VR Moscow, are highly resistant to verticillium wilt but not to fusarium wilt. Varieties highly resistant to both diseases are now being developed.

Losses from the leaf spot diseases and certain fruit rots caused by parasitic organisms can be reduced by the timely application of fungicides. They are particularly necessary in the Central, Atlantic, and Northeastern States, where a humid climate favours the occurrence of such diseases. The use of fungicides there has become much more general because of the continued appearance of late blight in sections where it was not a problem before 1945 and 1946. The routine use of fungicides also has been accelerated by the development of the very effective dithiocarbamate compounds. which have increased the possibility of reducing losses. Proper spraying and dusting and improved cultural methods have increased tomato yields markedly since 1940.

The fungicides most commonly used on tomatoes are copper compounds and organic compounds of the dithiocarbamate group. The neutral copper compounds (which include basic copper sulfates, copper oxychloride, copper oxychloride sulfate, and copper oxide) largely have supplanted bordeaux mixture because they are less injurious to tomato plants. As sprays, they are generally used in formulations that give the equivalent of 2 pounds of copper to 100 gallons of water. For seedbed spraying, however, this often is reduced to 1.5 pounds to 100 gallons. Bordeaux mixture cannot be safely used on tomato seedlings, but often is employed after the fruit is set.

An 8-4 100 mixture (in which the first form represents pounds of copper sulfate, the second hydrated lime, and the third water) is probably that most generally favoured, but 8-8 100, 6-6-100, and 6-3-100 preparations are used on tomatoes. Dusts containing the neutral Coppers are formulated so as to give 5 to 7 per cent of actual copper in a suitable inert carrier.

The dithiocarbamate compounds commonly used on tomatoes are zineb (zinc ethylene bisdithiocarbamate), nabam (sodium ethylene bisdithiocarbamate), which is used with zinc sulfate, and ziram (zinc dimethyl dithiocarbamate).

A newly introduced compound, manganese ethylene bisdithiocarbamate (Manzate), also appears to be effective against a number of tomato diseases. Zineb, ziram, and the manganese salt commonly are used at the rate of 2 pounds of the commercial preparation to 100 gallons of water. With nabam, 2 quarts of the commercial preparation are added to 100 gallons of water and then 1 pound of zinc sulfate is added.

Zineb and ziram also are used in dust preparations containing from 6 to 8 per cent of the fungicides. Other compounds such as Phygon (2, 3-dichloro-1, 4-naphthoquinone), captan (n-trichloromethylthio tetrahydrophalimide), Cop-O-Zink (tribasic copper sulfate and zinc sulfate), and Crag 658 (copper-zinc-chromate) also are being used. In many instances they have given satisfactory results but more experience with some of these materials is needed.

New organic compounds constantly are being tested for tomato disease control. Both the copper fungicides and compounds of the dithiocarbamate group are quite effective for the control of alternaria leaf spot. Late blight can be controlled with copper compounds and with those of the dithiocarbamate group other than ziram. Losses from gray leaf spot, caused by the fungus Stemphylium solani, and septoria leaf spot can be reduced by applying either copper or dithiocarbamate fungicides. Manganese ethylene bisdithiocarbamate appears to be very effective against gray leaf spot.

Ziram is particularly effective for control of anthracnose fruit rot but other dithiocarbamate compounds also will reduce losses from this disease. It is not satisfactorily controlled by copper fungicides. The relative extent of the use of copper and dithiocarbamate fungicides varies with the section in which tomatoes are grown. In southern

Florida, zineb and nabam have proved to be superior to the copper fungicides for late blight control and have been generally used. There has seemed to be some preference for the dithiocarbamate materials throughout the South Atlantic States, although copper fungicides also are recommended for use in tomato disease control. In the North Central, North Atlantic, and Northeastern States, copper fungicides are commonly used for control of late blight although the dithiocarbamate materials also have been extensively used. Seedbeds need protection from leaf spot diseases.

Often they are sprayed or dusted three or four times before the plants are taken to the field. In the field, the first applications of fungicides usually can be made about 30 days after the first flower cluster blooms. Later applications can be made at 10-day intervals unless there are frequent rains, when the interval may have to be reduced to 7 days, or even less, to secure good control.

Spraying has proved to be superior to dusting in the control of tomato diseases but, when wet soil makes it impossible to use heavy spraying equipment, dusting is a very valuable adjunct. The coverage obtained with ground dusting machines has been superior to that secured from applications of dust by airplane. We have no commercial varieties of tomatoes resistant to mosaic diseases and must depend on measures of sanitation to prevent infection.

After the diseases appear in the field or greenhouse, it is hard to check their further spread. Therefore the best means of control consists of preventing infection in the seedling plants and thus delaying the appearance of the disease until a crop is well started. If that can be done, the loss is likely to be less severe. One should never handle seedlings without first washing his hands with soap and water to remove any virus from the hands.

Because manufactured tobacco may carry the virus of tobacco mosaic, no one should smoke or chew tobacco while handling the plants. Workers can stop at intervals for a smoke and then wash their hands before again touching the plants. All perennial weeds around the seedbeds and in or near the fields should be destroyed because they may carry viruses of tobacco or cucumber mosaic. Every effort should be made to prevent aphid infestation of the seedbed by using insecticides regularly.

When seedlings are grown in a greenhouse used for commercial tomato production, the seedlings should not be handled after working with older plants unless the hands have first been washed. Anyone who has worked in a potato field or peeled or cut potatoes should wash the hands before working with tomatoes. The removal of mosaic plants in the field usually is of little value in checking the spread of mosaic diseases and often leads to further infection through brushing the infected plants against healthy ones in the course of removal. Because many nonparasitic diseases are induced wholly or partly by extremes of temperature or moisture, the grower is somewhat at the mercy of the elements where those disorders are concerned. In some diseases, however, both nutrition and weather have a part, and some losses may be avoided by care in the use of fertilizers, particularly nitrogen.

Chapter 7

Control of Plant Disease

Decay fungi are living organisms which send minute threads called "hyphae" through damp wood, taking their food from the wood as they grow. Gradually, the wood is decomposed and its strength is lost. Such damage is often inconspicuous until its final stages, and in a few instances homeowners have suddenly found floors breaking through or doors falling from their hinges due to wood rot.

When previously dry wood is placed in contact with moist soil, or in a location where it is subject to condensation (such as unventilated crawl space), it is likely that wood decay problems will occur. Rain leaks, faulty plumbing and leaky downspouts also are common sources of moisture. In some instances, water can be transported to the site of decay through strands or "rhizomorphs" of the decay fungi. Water-transporting strands may extend for thirty or more feet across brick, concrete or similar materials.

The wood decay fungus, *Serpula lacrimans*, has been known to transport water up three stories to an area where decay is occurring. *Poria incrassata* is also capable of transporting water long distances. However, these fungi are exceptions to the rule. Most wood-rotting fungi must have a direct supply of water at the site of decay. Thus the term "dry-rot," sometimes applied to decay in wood structures, is erroneous.

Prevention

- If the decay hazard is high, select the heartwood of decay-resistant species or use wood properly treated with a good preservative. Conifers from which decay-resistant lumber is produced include Pacific yew, juniper, redwood, baldcypress, and western red cedar. Durable hardwood species include osage orange, black locust, red mulberry, catalpa and black walnut.
- Build on a well-drained site. Use proper grading to prevent

water from seeping under the house. Install effective drain tile, roof overhang, gutters, and downspouts. Place no untreated wood within 18 inches of the ground.

- Provide adequate cross ventilation beneath buildings to eliminate dead air pockets. Install two square feet of opening for 25 linear feet of wall. Dense bushes or other plants should not be placed in front of these ventilators.
- Install a vapour barrier on the soil surface to cause soil moisture to condense on the barrier and return to the soil rather than condensing on the floor and above joists. Satisfactory barriers can be made by covering the soil with asphalt roofing paper or polyethylene sheets.

Repair of Decayed Buildings

First determine the source of moisture and remove it. If adequate ventilation and soil drainage are provided and all contacts of untreated wood with the soil or moist concrete or masonry are broken, decayed wood will dry out and further decay will be stopped. When making replacements, cut out at least one foot beyond the rotten area. Avoid placing new lumber in contact with old, decayed wood. Replacement lumber should be treated before installation. Remodel to provide more ventilation and better design rather than simply replacing decayed lumber.

BACTERIAL CROWN GALL OF FRUIT CROPS

Crown gall is caused by the bacterium *Agrobacterium tumefaciens*. This bacterium has the widest host range of any plant pathogen. It is capable of causing tumors, or "galls," on virtually all plant species, except the monocots (grasses). A similar bacterium, *Agrobacterium rubi*, causes galls on the canes of brambles. All fruit crops grown are susceptible. The disease is particularly destructive on brambles (raspberries and blackberries) and grapes. It can also cause severe problems on apple, pear, blueberry, all stone fruits and on ornamentals. The bacteria induce galls or tumors on the roots, crowns, trunks and canes of infected plants. These galls interfere with water and nutrient flow in the plants. Seriously infected plants may become weakened, stunted and unproductive.

Symptoms

The disease first appears as small overgrowths or galls on the roots, crown, trunk or canes. Galls usually develop on the crown or trunk of the plant near the soil line or underground on the roots. Above ground or aerial galls may form on canes of brambles and highly susceptible cultivars of grape. Although they can occur, aerial galls are not common on fruit trees. In early stages of development the galls appear as tumor-like swellings that are more or less spherical, white or flesh-coloured, rough, spongy (soft) and wart-like.

They usually form in late spring or early summer and can be formed each season. As galls age they become dark brown to black, hard, rough, and woody. Some disintegrate with time and others may remain for the life of the plant. The tops of infected plants may appear normal. If infection is severe, plants may be stunted, produce dry, poorly-developed fruit, or show various deficiency symptoms due to impaired uptake and transport of nutrients and water.

Causal Organism

The crown gall bacterium is soil-borne and persists for long periods of time in the soil in plant debris. It requires a fresh wound in order to infect and initiate gall formation. Wounds that commonly serve as infection sites are those made during pruning, machinery operations, freezing injury, growth cracks, soil insects and any other factor that causes injury to plant tissues.

Bacteria are abundant in the outer portions of primary galls, which is often sloughed off into the soil. In addition to primary galls, secondary galls may also form around other wounds and on other portions of the plant in the absence of the bacterium. The bacteria overwinter inside the plant (systemically) in galls, or in the soil. When they come in contact with wounded tissue of a susceptible host, they enter the plant and induce gall formation, thus completing the disease cycle. The bacteria are most commonly introduced into a planting site on or in planting material.

Control

- Obtain clean (disease free) nursery stock from a reputable nursery and inspect the roots and crowns yourself to make sure they are free from galls. Avoid planting clean material in sites previously infested with the bacteria.
- Avoid all unnecessary root, crown and trunk wounding by careless cultivation and other machinery operation, and control soil insects. Any practice that reduces wounding is highly beneficial. Preventing winter injury (especially on grapes) is also beneficial.
- On grapes, the double trunk system of training may be a useful system for minimizing losses due to crown gall. If one

trunk is infected, it can be removed. The remaining trunk can be pruned leaving a full number of buds until the second trunk can be renewed. Galls on the upper parts of the trunk or on canes can be removed by pruning.

- A relatively new biological control agent for crown gall is available for apple, pear, stone fruit, blueberry, brambles and many ornamentals. It is not effective on grape. The agent is a nonpathogenic strain of bacterium (*Agrobacterium radiobacter* strain 84) that protects the plants against infection by the naturally occurring strains of pathogenic bacteria in the soil. Nursery stock is dipped in a suspension of commercially prepared *Agrobacterium radiobacter* strain 84 at planting time. The antagonistic bacteria act only to protect disease free plants from future infection by the crown gall bacterium; they cannot cure infected plants.

BITTER ROT OF APPLES

Bitter rot is a common disease of apples and pears in practically all countries where they are commercially grown. Of the three fruit rot diseases on apple (bitter rot, white rot and black rot), bitter rot has the potential to be the most destructive. The fungus that causes bitter rot (fruit rot) can also cause a leaf spot and canker, although the leaf spot and canker form of the disease are not common. The disease is most common in warmer regions because high temperatures favour disease development. Before the development of effective fungicides, entire crops were lost to bitter rot during periods of warm, wet weather.

Symptoms

Fruit rot symptoms differ somewhat, depending on whether infection is initiated by spores from perithecial strains of the fungus (strains that produce ascospores and conidia) or conidial strains of the fungus (strains that produce only conidia). Initial symptoms produced by perithecial or conidial strains are similar. Lesions begin as small, slightly sunken areas, which are light brown to dark brown.

On mature fruit, lesions may be surrounded by a red halo. Lesions originating from infections by conidial strains remain circular and become sunken as they enlarge. Copious quantities of ooze containing conidia (spores) are produced in fungal fruiting bodies called acervuli, which occur in concentric circles on rotted tissue around the point of infection. Acervuli are sparse on some lesions and very dense on others. Under moist, humid conditions, the spore masses appear creamy and are salmon to pink in colour. Lesions initiated by perithecial strains are usually not sunken and are often darker brown than those caused by conidial strains. Acervuli are widely scattered over the surface, and perithecia are found in dark brown to black clumps scattered on the surface. Bitter rot lesions on all fruit (regardless of which strain caused infection) extend in a cone shape manner towards the core. In cross section, the lesion appears "t-shaped".

This is a reasonably reliable characteristic that can be used to separate bitter rot from white rot or black rot. The rotten area is brown but much firmer than white rot. Infected fruit eventually mummify, and some may remain attached to the tree through the winter. Leaf lesions are not common and are caused by the perithecial strain. They begin as small, red flecks, which enlarge to irregular brown spots 1/16 to 1/2 inch in diameter. Severely affected leaves may drop off the tree. Bitter rot cankers are rare in the eastern United States. Cankers are oval, sunken and often zonate in appearance.

Casual Organism and Disease Cycle

Bitter rot is caused by the fungus, *Glomerella cingulata*. The fungus overwinters in apple orchards in dead wood or mummified fruit in the trees that were infected during the previous season. Conidia, produced in these overwintering sites, are the primary inoculum source in the spring, although ascospore inoculum is important in some orchards. Conidia are spread by splashing and wind-blown rain. Insects and birds are also involved in their dispersal. Ascospores are released after rain and are airborne. Fruit are susceptible to infection from 3 weeks after petal fall until harvest. Temperatures of 80 or 90 degrees F are most favourable for disease development. Because of the large number of conidia produced in lesions on fruit, the fungus has the potential for rapid spread within the orchard.

Control

- Control of bitter rot is best achieved through an integrated programme of cultural practices and chemical control measures.
- Sanitation is critical for effective control. Piles of prunings are an important source of inoculum and should be removed from the perimeter of the orchard or burned. Prunings can be left on the orchard floor if they are chopped with a flail mower, which removes much of the bark and allows them to decompose faster.
- Removal of mummified apples and pruning out dead wood in the tree are important for reducing the inoculum within

the tree. Pruning out current-season shoots infected with fire blight is also important, because they can be colonized and serve as an inoculum source during the same growing season.

- Any practice that helps to maintain trees in a healthy vigorous condition is critical for controlling the canker phase of the disease. Cankers generally develop only on stressed or weakened trees. Prune trees annually and maintain a balanced fertility programme based on soil and foliar nutrient analysis. Cankers generally develop rapidly on winter-injured trees.
- The use of fungicides combined with good sanitation is beneficial for controlling the fruit rot phase of the disease. Fungicides are not effective for controlling the canker phase of the disease on weakened trees.

Wild Mushrooms

There are 2,000 or more kinds of wild mushrooms. Some are poisonous and some are edible and delicious when properly prepared. The edibility of the majority is either not known or they are not considered for food because of their small size or poor flavour or texture. Even though not every one is interested in collecting mushrooms to eat, it is important to understand most have an important and beneficial role in the environment.

They grow in a wide variety of habitats. Most of the mushrooms seen on a walk through a woods are beneficial. Many species are quite specific about their food source and will be found only under or near certain kinds of trees-some under pines, others under oak, etc. Some are important as decay organisms, aiding in the breakdown of logs, leaves, stems and other organic debris.

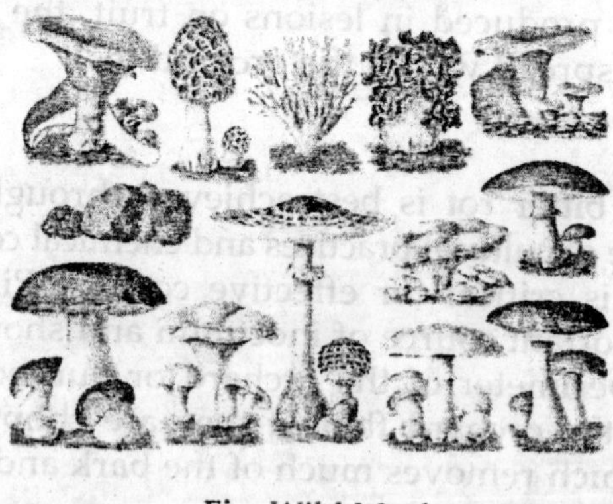

Fig. Wild Mushrooms

This important role of mushrooms results in recycling of essential nutrients. Some mushrooms grow in, and form their fruiting structures on living trees causing decay of the sapwood or of the heartwood. Many woodland mushrooms are essential to good growth, and even survival of trees. They establish a relationship with roots of living trees that is mutually beneficial. These are called mycorrhizal mushrooms. All mushrooms, whether poisonous or edible can be admired for their beauty and the fantastic variety of form, colour and texture.

WHICH MUSHROOMS ARE SAFE TO EAT?

Some edible mushrooms are very similar in appearance to poisonous kinds and may grow in the same habitat. Edible mushrooms are known to be safe to eat because they have been eaten frequently with no ill effects. Poisonous mushrooms are known because someone ate them and became ill or died. There is no test or characteristic to distinguish edible from poisonous mushrooms. This indicates a need to identify with certainty one of several of the proven edible species and pick and eat only those positively identified. At the same time, you should also learn to identify some of the common poisonous mushrooms, especially those that are similar to edible kinds.

It is especially important to learn the characteristics of the Amanita mushrooms, since several of the species common are poisonous, a few causing serious illness and sometimes death. The word "toadstool" is often used to indicate a poisonous mushroom. Since there is no way to distinguish between a so-called "toadstool" and an edible mushroom it is more precise to speak of poisonous mushrooms or edible mushrooms. The season for collecting wild mushrooms for food begins in late March and early April when the first morel or sponge mushrooms are found. These choice edible mushrooms are most abundant during April and the first two weeks of May. The false morels (members of the Gyromitra genus) are found at this same time of the year, but they must be regarded as poisonous and not collected for eating.

It is true that many have eaten false morels with no apparent ill effects. However, recent research has shown toxins to be present in some of the false morels that can cause death or serious illness. Do not eat the false morels. From mid summer to late autumn, a great variety of mushrooms may be found. A number of these are choice edibles. Photographs and brief descriptions of several of the more common mushrooms found are included in this fact sheet.

Control of Nuisance and Detrimental Molds (Fungi) in Mulches and Composts

Mulches and composts are often used to improve soils and plant health and to control weeds. They improve drainage as they decompose even though the ability of the soil to hold moisture is increased. They lower soil temperature in the summer and insulate roots from cold in winter conditions.

Eventually, they mineralize, release nutrients for plants, and leave humic substances as residues. Their beneficial side effects gradually disappear unless more mulch or compost is applied. Generally, these organic materials inhibit undesirable microorganisms such as soilborne pathogens that cause diseases of plants.

They also stimulate the activity of many types of beneficial microorganisms, including mycorrhizal fungi. Occasionally, however, microorganisms (primarily fungi) in mulches and composts can become a nuisance and even cause certain diseases of plants. Whether a mulch or a compost provides beneficial or detrimental effects is largely determined by the type of organic matter from which it was produced and the degree to which it was decomposed and treated before its application in the landscape. The temperature, pH, and moisture content of the products just before application also have an effect. The severity of nuisance fungi can be minimized if appropriate steps are taken in time.

Examples of Nuisance Fungi

The shotgun or artillery fungus (*Sphaerobolus*) may cause serious problems. While it decays the mulch, it also produces fruiting structures that resemble tiny cream or orange-brown cups that hold a spore mass resembling a tiny black egg (1/10 inch in diameter). This fungus shoots these spore masses high into the air. They stick to any surface and resemble small tar spots on leaves of plants or the siding of homes. They are difficult to remove, leave stained surfaces, and may result in major damage.

Slime molds are another type of nuisance fungus. They first appear as bright yellow or orange slimy masses that may be several inches to a foot or more across. They produce tiny spores that eventually dry out and blow away. These molds, like many others such as stink horns and bird¹s nest fungi, actually should be considered microbial ornamentals in the landscape. However, some fungi in mulches and composts produce toad stools (mushrooms), and some of these are toxic to humans. It is a good idea to destroy them when small children have access to the mulched area.

Another fungal problem that is often not identified correctly occurs when mulches are applied too deep (4-6 inches) instead of the ideal depth of 1.5 to 2 inches. Deep layers of mulch, particularly if prepared from fresh woody materials, may actually undergo high temperature decomposition during the summer. The result is that the mulch dries out to less than 34 per cent moisture and becomes a dusty mass. Fungi often colonize these dry mulches until they become a water-repelling, moldy chunk of material. Young trees mulched in this way sometimes die from drought even though the homeowner irrigates the area, because water runs off the mulch since it repels water.

Other Types of Problems

Fresh mulches prepared from trees killed by plant diseases may be colonized by plant pathogens. *Verticillium dahliae,* a fungus that causes wilts and death of many shade trees and ornamental shrubs, can be carried in infested mulch and kill susceptible plants in the landscape. *Rhizoctonia solani,* another plant pathogen that causes damping-off of many types of seedling plants, is actually stimulated by fresh mulches.

This pathogen utilizes the cellulose in wood as a source of food. Short-term composting of mulches in windrows under high-temperature conditions (130-160 degrees F) kills these plant pathogens. Six weeks of composting is sufficient to kill most plant pathogens and avoid their dissemination in mulches or composts.

Mycorrhizae, which are fungi that can form beneficial associations with roots, also are affected by mulches and composts. A shallow layer of wood chips (1-2") or compost improves tree establishment because mycorrhizae are stimulated by the slow release of organic sources of nitrogen and carbon in organic matter.

However, a deep layer (4-6") of the same freshly chipped wood has been shown to inhibit the development of mycorrhizae during reforestation. Negative effects on mycorrhizae must be avoided in the landscape because they are very important in the maintenance of healthy plants.

Compost and mulch producers, landscapers, and homeowners can take measures to minimize fungal problems in the landscape. The type of mulch used, fresh versus composted mulch, the moisture content of the mulch before and during its utilization, the temperature and pH of the mulch before and during utilization, and the depth to which it is applied all play a role. Each factor is discussed here.

MULCH TYPE AND FRESH VERSUS COMPOSTED MULCH

Wood products from some trees are more resistant to decay than others and, therefore, cause fewer problems. Bark chips (nuggets) from large mature pine or other softwood trees such as cypress trees contain mostly lignin (dark material in bark), wax and protected cellulose that resist decay. On the other hand, wood wastes from these same tree species, but ground as young trees, rot quite readily because the cellulose in such bark and wood products is not yet protected from decomposition by lignin waxes or tannins.

Hardwood tree bark (oak, maple, etc.), even from large trees, contains a large concentration of cellulose that is not protected from rotting. Therefore, hardwood bark mulches, like ground wood from almost all tree species, rot readily and cause most of the nuisance mold problems in the landscape. The finer the product is ground, the more severe the problem can be! These materials are low in nitrogen content.

The fine particles (less than 3/4" diameter) in such mulches cause nitrogen immobilization in soil. The microflora that decomposes the wood particles takes up the nitrogen required for growth of plants. The result is that the plant becomes starved for nitrogen. Some mulch producers screen all particles smaller than 3/8" out of high-wood-content or hardwood-bark mulches, which avoids most of the nitrogen immobilization problem.

The best way to avoid all these problems and bring about beneficial effects by mulching is to add nitrogen to woody and hardwood bark products followed by composting to lower the carbon to nitrogen ratio. Blending of grass clippings with wood wastes before composting is one way to achieve this. Addition of poultry manure or urea to supply 1.2 lbs. available nitrogen per cubic yard of material satisfies the nitrogen need also. Some landscapers add 10-15 per cent by volume composted sewage sludge to hardwood bark or wood wastes, and this makes an ideal product that has performed very well in landscapes.

These amended products should be composted at least six weeks. This process kills plant pathogens, eggs of insect pests, and produces a nitrified product that releases plant nutrients rather than ties up nitrogen. As mentioned above, the microorganisms that have colonized these products reduce the potential for growth of nuisance fungi and provide control of many plant diseases.

TEMPERATURE, MOISTURE CONTENT, AND pH

Landscapers often apply quality mulch products from high temperature piles (140-160 degrees F) directly into the landscape. The temperature of the mulch is high because of heat produced by growth of microorganisms, known as thermophiles, in storage piles during the composting process. These microorganisms die soon after the mulch cools to 50-80of after it has been placed around homes. Because they require high temperatures to survive, they cannot grow and compete with soil microorganisms at the low temperature of mulches in the landscape.

The sudden temperature drop that often occurs after mulch is applied creates what is known as a "biological vacuum." It also can occur during bagging of products at producers of mulches and particularly during dry seasons. Mesophiles (low temperature soil microorganisms) rapidly colonize such mulches.

If the mulch is dry, or dries out to a moisture content below 34 per cent during the first day after it is applied (mulches are dusty below this moisture content), fungi become the primary colonizers. This sets the stage for problems later, and the problem becomes most severe in mulches that are applied too deep in the landscape.

After prolonged heavy rains, the dry material colonized by fungi eventually becomes wet. Dry products stored in bags may also become moist when water produced as a result of microbial activity accumulates along the inner surface in bags. Bacteria then rapidly colonize the fungal white mass to induce the formation of fruiting structures by the fungi. The nuisance toad stools and other fruiting structures appear a few days later. Mold problems occur also when dry products are bagged or applied to dry soils.

Dry composts removed from high-temperature piles occasionally cause mushroom problems in bags and also in soils at nurseries. These moldy products inhibit plant growth in field soils as well as in potting mixes. They also cause wettability problems if dry conditions persist in the soil for a few weeks to give fungi a chance to become the dominant colonizers. Plants do not grow well in such moldy soils. These problems can be reduced by soaking the high-temperature products with water as they are applied in the landscape or bagged. The high-moisture organic matter then becomes rapidly colonized by bacteria during the first few days.

These bacteria compete with fungi to reduce the potential for the development of major mold problems. This strategy has been successfully applied over the past decade to hardwood as well as softwood composts and mulches. It has controlled nuisance problems caused by many fungi in various parts of the United States and abroad.

The pH or acidity of the mulch is another important factor. Sour mulches that give off acrid odors may range in pH from as low as 2.5 up to 4.8. Highly acidic mulches are toxic to most plants and promote the growth of fungi. Bacteria that inhibit fungal growth cannot colonize mulches when the pH is lower than 5.2. The low pH and fungal problems are avoided if the raw material is nitrified and composted as described earlier. In summary, water applied at the right time during composting, storage, and mulching can solve most of the fungal nuisance problems. It is best to maintain a water content higher than 40 per cent on a total weight basis. Again this allows bacteria as well as fungi to colonize the organic matter, and it sets up competition for nuisance molds. The moisture content of most organic products actually can be raised above 50 per cent and not present excessive weight problems during transport.

WHAT TO DO ONCE THE PROBLEMS OCCUR

Sometimes very little can be done to control nuisance fungi other than to spade the mulch into the surface soil layer followed by soaking with water. Another option is to remove the mulch, place it in a heap after thorough wetting to allow for self-heating to occur (110-140 degrees F). This will kill nuisance fungi. If fresh dry mulch is placed on top of mulch colonized by nuisance fungi, the problems may occur again the following year or even earlier.

The best control strategy for homeowners and landscapers is to purchase composted products low in wood content. Fresh, finely ground woody products should be avoided for many reasons unless composted first. Coarse fresh woody products are much less likely to cause problems unless applied too deep. It is important to soak all mulches immediately after they have been applied.

Generally, mulches should not be applied to a depth greater than two inches. Mulches and composts applied in this manner provide many types of beneficial effects rather than nuisance problems, or worse, plant diseases. Sour mulches should be avoided altogether.

MYCORRHIZAS IN THE URBAN LANDSCAPE

Mycorrhizas are anatomically intimate associations between fine (feeder) roots of plants and some special soilborne fungi. For this reason they are called symbioses. The plant, being capable of producing its own food through photosynthesis, can potentially survive without the fungus. The latter, however, is often completely dependent on the live plant for sustenance; thus, it is a parasite of the plant.

The association usually (but not always) results in a nutritional benefit to both the host plant and the fungus: the plant receives extra soil mineral nutrients, such as phosphorus and nitrogen, and extra water absorbed by the fungus growing out into the soil; the fungus obtains sugars and other vital substances from the plant. This kind of two-way beneficial interaction is called a "mutualistic association."

In nature, the vast majority of plants are affected by this kind of association with fungi, to the point that some scientists claim that "the study of plants without their mycorrhizas is the study of artefacts. The majority of plants, strictly speaking, do not have roots, they have mycorrhizas".

In fact, it is now widely accepted that in the course of evolution, colonization of land by plants occurred thanks to these plant-fungus associations. Mycorrhizas are important because plants grown in the absence of their fungal associates generally do not perform as well as those that are mycorrhizal. Sometimes, such as in the case of orchids, the seeds do not even germinate in the absence of the appropriate fungus.

Mycorrhizas are also known to protect the plants from various root diseases. However, lack of mycorrhizal roots on normally mycorrhizal plants is a rather rare occurrence, as most natural soils contain a full complement of mycorrhizal fungal propagules, that is, spores or other structures used by the fungi to infect new plants.

Mycorrhizas Come in Different Types

There are seven different kinds of mycorrhizas. The most important ones from an urban landscape perspective are the ectomycorrhizas, the endomycorrhizas, and the ericoid mycorrhizas. The main and most visible characteristic of ectomycorrhizas is the fungus growing on the surface (hence "ecto," which means "outside") of the feeder roots. This fungal growth is called a sheath or mantle. The fungus grows from the mantle into the cortical layers of the root, but remains between the plant cells and does not penetrate them, forming a so-called Hartig net.

This is where the nutrient exchanges between the plant and the fungus occur. On the other hand, endomycorrhizas are characterized by the absence of a mantle and by penetration of the fungus into the cortical cells. Once inside the cells, the fungus produces characteristic structures, called "arbuscules" (= tiny trees) and "vesicles."

The arbuscules are the structures where the nutrient exchanges between the plant and the fungus occur. Hence, endomycorrhizas are also called "vesicular arbuscular mycorrhizas" (VAM) or just "arbuscular mycorrhizas" (AM). Ericoid mycorrhizas are somewhat similar to VAM, in that the fungus invades the host cells, but affect only plants in the family Ericaceae.

Different plant groups tend to form specific types of associations. For example, Ericaceae tend to form ericoid mycorrhizas, P inaceae almost always form ectomycorrhizas, and hardwoods can form both ecto- and endomycorrhizas, sometimes on the same plant. Table shows a partial list of forest and ornamental trees and their specific type of association.

One last characteristic to bear in mind, particularly when considering the following paragraph, is that the fungi involved in the two main types of mycorrhizas, that is endo- and ectomyco-rrhizas, feature remarkably different levels of host species specificity. While most individual ectomycorrhizal fungal species tend to form associations with specific host plant species or a restricted number of them, individual endomycorrhizal fungal species are "generalists" and can associate with hundreds of different host plant species. This characteristic is reflected in the number of recognized mycorrhizal fungal species: thousands of ectomycorrhizal vs. about 200 endomycorrhizal fungal species.

ARTIFICIAL INOCULATION

This question is becoming more and more common as more companies have recently begun commercializing mycorrhizal inoculum. Artificial mycorrhizal inoculation may benefit plants only when there is no natural and appropriate mycorrhizal inoculum in the soil, or the inoculum level is low, or species present are less efficient at aiding the plant host than those being introduced.

This situation is most commonly found in highly impoverished soils that do not support natural fungal growth, for example in the case of mine spoils. In principle, highly compacted, organic nutrient-poor soils in urban environments (often the result of mineral subsoil horizons having been turned into the top layers by construction activities) may also be prime candidates for artificial mycorrhizal inoculation.

However, success of the treatment can never be guaranteed for two main reasons:

- The urban soil may already contain a complement of mycorrhizal fungi, however depleted, that out-compete the newly introduced species. Competition among mycorrhizal fungi is well documented and a common occurrence in natural soils, for example in reforestation operations. It is particularly important for ectomycorrhizal fungi;
- The soil organic matter content may not be adequate. The presence of organic matter can greatly enhance the development of mycorrhizas.

Due to the specificity properties of the different mycorrhizal associations described above, it is likely that mycorrhizal inoculation will be more successful, in terms of actual establishment of the inoculated species, with endomycorrhizal fungal species, even though the main ectomycorrhizal inoculum being commercialized is a "generalist.

Among specialists." At present, there is very limited, unbiased scientific evidence that artificial mycorrhizal inoculum in the landscape makes plant establishment more successful or that the inoculated plants grow better over time. In fact, the available evidence is rather inconsistent. In part this is due to the fact that commercial preparations may not be very viable or not viable at all by the time they reach the consumer, or become not viable once they are in the consumer's hands.

But perhaps the biggest factor is that soil conditions in the landscape may be too variable to be able to make generalizations about the usefulness of artificial inoculation with the few selected species available commercially.

Furthermore, any short-term positive effects observed on plant growth following mycorrhizal inoculation may simply be the result of the fertilizer components that are often present in the commercial preparations. The bottom line is that artificial inoculation may have either a positive or neutral effect on plant growth and health. In all cases, however, it is very unlikely that artificial inoculation is detrimental to the plant.

In other words, the treatment may or may not work, but would not be injurious. Current research, conducted in the author's and other researchers' laboratories, suggests that organic soil amendments, particularly the addition of composted mulches, greatly enhances the mycorrhizal status of landscape plants, even without the addition of artificial inoculum. Improvement of soil conditions may therefore be the most important factor in mycorrhizal development in the urban landscape.

OAK WILT

Oak wilt is a serious and often deadly vascular disease of oaks. The fungal pathogen, Ceratocystis fagacearum, is believed to be native to the United States and is distributed throughout the Midwest and Texas.

What the Pathogen Does

The fungus grows into and throughout the water conductive tissues (that is, the sapwood) of the host. The fungus plugs the vessels with its own body (mycelium) and spores, but it also causes a defensive reaction by the tree to stop the fungal spread by actively plugging its own vessels. These processes interfere with water uptake and cause a wilting syndrome which often results in death of the tree.

Susceptible Oaks

All oaks are susceptible. Those in the red-black oak group (black, blackjack, pin, northern and southern red, scarlet, shingle and shumard oak) are extremely susceptible and can die within a few weeks of infection. Oaks in the white group (bur, chinquapin, post, swamp white, and white oak) are more tolerant of the disease and may survive infection for one or more years while displaying decline symptoms.

Diagnostic Symptoms

Symptoms are typical of wilts. Leaves usually begin withering in the upper canopy, producing "flags," that is, whole branches or crown portions turning red-brown. Leaves of red oaks typically show yellowing and browning of the leaf margins. White oak leaves usually show rather non-descript symptoms. Conversely, live oaks in the southern United States produce characteristic dead areas along the leaf veins.

These dead areas generally expand until the whole leaf becomes brown. Eventually the leaves fall from the tree. If infections occur in late spring, trees usually begin wilting in mid-summer to late summer, when the plants often are subjected to water deficit due to increased transpiration demand and decreased rainfall.

A specific and sufficient diagnostic character is the appearance on dead and dying red oaks (but not white oaks) of spore-bearing fungal mats under the desiccating bark. These fungal mats crack the bark open with pressure pads to facilitate dissemination of the pathogen. Sapwood streaking is also a good, but insufficient, diagnostic character.

Disease Cycle and Conditions Favouring Disease

In order to properly manage oak wilt it is essential to understand its cycle. The pathogen spreads from diseased to healthy trees in two ways: overland and underground. Overland spread is mediated mainly by sap feeding (a.k.a. picnic) beetles (Coleoptera: Nitidulidae). However, there is some evidence that oak bark beetles (Coleoptera: Scolytidae) may also be involved. Nitidulids are attracted by chemicals emanating from the fungal mats described above.

Once on the mats, the beetles pick up fungal spores and can carry them, sometimes over distances of a few miles, to freshly wounded healthy trees (attracted by the smell of fresh sap). This results in new infections, thus closing the overland cycle. While insect spread is an important medium to long range dispersal mechanism for this fungus, it is estimated that 90 per cent of new infections occur between neighbouring trees through root grafts.

In this case, the fungus grows down the trunk, into the roots of diseased trees, and then into healthy trees via the common root system. Once in the new tree the pathogen grows throughout the vascular system and spreads to other trees via the root system or the beetles. In this way, spread through root systems often results in disease centres that expand outward from the initially infected tree. From the above, it follows that conditions favouring disease include the availability of susceptible oak species, trees growing close to each other, and the availability of fresh wounds for beetle-mediated infection.

Pruning wounds are obvious culprits, but any fresh wound will function as potential infection gateway. The word fresh is emphasized because it is believed that wounds are attractive to Nitidulid beetles only for up to three days. As with many plant diseases, other stresses (for example drought) can predispose trees to faster symptom development, and thus worsen the syndrome.

Control and Management of the Disease

The best control for oak wilt is through preventative measures that interrupt the disease cycle.

Prevention of Overland Spread

Overland spread can be hindered or interrupted by ensuring that trees are never wounded between April 15 and July 1. This is when most Nitidulid beetles fly to locate fresh sap and/or fungal mats. A more stringent approach is to avoid wounding the trees throughout the growing season, since additional summer flights of the beetles are possible.

If pruning is absolutely necessary during the growing season, it is imperative to dress the wounds. This can be done with latex paint. Although this will slow wound healing, it will also deter beetles from landing on the wounds.

Prevention of Underground Spread

Given the higher significance of underground spread, control of direct tree-to-tree transmission is much more important. Here, interruption of the disease cycle is accomplished by physically severing actual or potential root contacts between diseased and healthy trees.

This is done by trenching or cutting through the soil with a trencher or vibratory plow. The latter is the preferred tool. Given the depth of oak root systems, it is advisable to use a 5 ft blade. Trenching must always be done before the diseased or dead trees are cut for removal, to avoid sudden water tension imbalances that might "suck" fungal material from the infected trees into the healthy trees through the common root system.

Trenching should be conducted by advice of specialists. This is due to the importance of locating the trenches appropriately between diseased and healthy trees. When possible, a double trench defining a buffer band of apparently healthy trees between diseased and uninfected trees should be used. On residential or commercial properties, always determine the location of buried utility lines which may affect the ability to completely sever the graft.

Furthermore, walkways, paths, and roads must also be considered appropriately, as tree roots commonly grow under them. Due to all of these potential obstacles to proper trenching, it is advisable to undertake such operations under the supervision of tree care professionals with expertise in the management of oak wilt.

There is currently no evidence that the blade will spread the pathogen. However, it is good precautionary practice to spray the blade to runoff, between trenches and between plots, with an antiseptic such as Lysol or a 20 per cent bleach solution.

Disposal of Dead and Dying Trees

Once the trenches are in place, diseased and dead trees should be removed as soon as possible by cutting them down to leave a 2-4 inch high stump. Because diseased trees with bark tightly attached may produce or harbor fungal mats, they should be disposed of promptly. Once the bark becomes loose or sloughs off, no mats can be produced and movement of the infected wood out of the diseased area is no longer a concern in an urban context. (However, strict restrictions apply for movement of diseased wood out of state and internationally.) Thus, either the trees are debarked mechanically, or the timber can be sold to a sawmill for cutting or chipping.

Although no studies are known on the transmissibility of C. fagacearum via wood chips, the pathogen does not survive when exposed to desiccation and is very temperature sensitive. Composting the chips would further reduce or eliminate the pathogen. Thus, it is highly unlikely that wood chips will spread the disease. However, good precautionary practice suggests to avoid using infested chip mulch around healthy oaks. If the wood cannot be disposed of as described above, it can be cut and split for firewood. Because this process does not involve debarking, firewood can still potentially harbor fungal mats and thus attract Nitidulids during the summer in which the trees died.

The wood must be arranged in stacks and covered with 4 mil plastic tarp through the winter (if the wood is used then) or the end of the next season (Oct. 1 of the year following the death of the trees). By producing a greenhouse effect, tarping will kill the temperature-sensitive pathogen and prevent the beetles from accessing potential fungal mats. Tarping should be done with transparent plastic to produce the desired greenhouse effect. However, black plastic will also work, by concentrating the sun's heat. In both cases, the best results are achieved by placing the tarped pile in an un-shaded, possibly sunny area.

When covering the pile, the tarp should be sealed to the ground to prevent beetles from accessing the pile. For this reason, all punctures in the tarp should be mended with duct tape. At the end of the second season the wood can be safely uncovered and disposed of as preferred, since it no longer constitutes a threat.

Chemical Treatments

Chemical treatments are usually not warranted, due to the high cost of intervention. However, application of systemic fungicides is an option when highly valuable trees are threatened by infected neighbouring trees, or whenever a high risk of infection exists. High value may be attributed to individual trees or groups of trees in communities, or it may apply to individual trees in a homeowner's yard.

Systemic fungicides have been demonstrated to be effective, particularly when applied as a preventative treatment. The only scientifically tested systemic fungicide showing any efficacy and labeled for use against oak wilt is propiconazole, available under the trade name Alamo.If a decision is made to apply propiconazole, it must be done strictly according to label and should be carried out by experienced, professional tree service personnel. The product is injected directly into the sapwood on root flares just under the soil line.

This is the best guarantee that the fungicide will be translocated throughout a tree, thus affording the maximum possible protection. In consideration of the disease cycle, the best time of the year to inject trees is early spring. However, application should occur as soon as the risk to a tree is realised, even if it is later in the growing season.

Depending on the tree size and value, treatments should be applied every 12-36 months, with annual assessments. This treatment has virtually no hope of succeeding in infected red oaks, even in early infection stages.

Chemical treatment has a higher chance of success, but only as a palliative measure, with the more tolerant white oaks. In this group, application of the fungicide to trees in early infection stages can result in delay of symptoms and eventual death. It will not, however, rid an infected tree of the pathogen.

ANTHRACNOSE FRUIT ROT OF PEPPER

Several species of plant pathogenic fungi in the genus *Colletotrichum* cause anthracnose in peppers and many other vegetables and fruits. Until the late 1990s, anthracnose of peppers and tomatoes was only associated with ripe or ripening fruit. Since that time, a more aggressive form of the disease has become established. This form attacks peppers at any stage of fruit development and may threaten the profitability of pepper crops in areas where it becomes established. This disease can also affect tomatoes, strawberries, and possibly other fruit and vegetable crops.

Symptoms

Circular or angular sunken lesions develop on immature fruit of any size. Often multiple lesions form on individual fruit. When disease is severe, lesions may coalesce. Often pink to orange masses of fungal spores form in concentric rings on the surface of the lesions. In older lesions, black structures called acervuli may be observed. With a hand lens, these look like small black dots; under a microscope they look like tufts of tiny black hairs.

The pathogen forms spores quickly and profusely and can spread rapidly throughout a pepper crop, resulting in up to 100 per cent yield loss. Lesions may also appear on stems and leaves as irregularly shaped brown spots with dark brown edges.

Causal Organism and Disease Cycle

This form of pepper anthracnose is caused by the fungus *Colletotrichum acutatum*. The pathogen survives on plant debris from infected crops and on other susceptible plant species. The fungus is not soil-borne for long periods in the absence of infested plant debris. The fungus may also be introduced into a crop on infested seed. During warm and wet periods, spores are splashed by rain or irrigation water from diseased to healthy fruit. Diseased fruit act as

a source of inoculum, allowing the disease to spread from plant to plant within the field.

- Plant only seed from disease-free plants or seed treated to reduce any fungal populations. Seed can be disinfested with a 30-minute soak at 52°C. for complete seed-treatment instructions.
- Peppers should be rotated out of infested fields for at least three years. Fields should be planted with crops other than tomatoes, eggplants, or other solanaceous crops or strawberries, which are also hosts of *C. acutatum*.
- Apply overhead irrigation during the early part of the day so that plants can dry before sundown.
- In areas where market constraints and other diseases do not limit the choice of cultivar, cultivars demonstrating a moderate level of resistance should be chosen.
- Apply protective fungicides at flowering when environmental conditions are favourable for disease development. Rates and timing of all further applications should be done according to product label instructions.

Index

S

T

V

W
